PowerCLI Essentials

Simplify and automate server administration tasks with PowerCLI

Chris Halverson

BIRMINGHAM - MUMBAI

PowerCLI Essentials

First published: April 2016

Production reference: 1210416

Published by Packt Publishing Ltd.
Livery Place
35 Livery Street
Birmingham B3 2PB, UK.

ISBN 978-1-78588-177-0

www.packtpub.com

Credits

Author
Chris Halverson

Reviewer
Tom Franken

Acquisition Editor
Divya Poojari

Content Development Editor
Onkar Wani

Technical Editor
Menza Mathew

Copy Editor
Merilyn Pereira

Project Coordinator
Bijal Patel

Proofreader
Safis Editing

Indexer
Mariammal Chettiyar

Graphics
Disha Haria

Production Coordinator
Arvindkumar Gupta

Cover Work
Arvindkumar Gupta

About the Author

Chris Halverson is a senior consultant for VMware in the Professional Services Organization in Canada. He specializes in the full Software-Defined Data Center (SDDC) stack, architecting, designing, and deploying customer solutions. He has been active in the VMware community as a VMUG leader for the past 4 years and prides himself as an active participant in the local IT market over the past 17 years. Previous roles have allowed him to work on enterprise architecture bringing process and rigor to the administration aspect of the position and automation that made the job smarter. There is even an aspect where "I replaced myself with a small script" has been heard from him when discussing the former roles.

When Chris is not working on building the virtual community, he shares his time and energy with his tremendous wife, three boys, and one princess. Excited to spend time watching and helping them figure out life for themselves, he encourages them in their sports, through their schooling, and through their own walk of faith.

Over the past few years, Chris has also been able to achieve a dream as a second degree black belt in Tae Kwon Do, crossing off one of those bucket list items.

Chris has also acted as a book reviewer.

I want to acknowledge my family, as they supported me through this time-consuming venture. There were many times when I had to forgo doing something with them to write this book.

Also, I'd like to shout out to my good friend, Adam Wysockyj, who was instrumental in getting me through the NSX portion of the book.

And last but not least, I want to thank God for the strength to write this. There were many days when I felt exhausted and through prayer was sustained and motivated to persevere.

About the Reviewer

Tom Franken has been supporting IT infrastructures for 20 years and VMWare ESX since the 2.5 days. He has extensive experience of using PowerShell and PowerCLI to automate processes and free his weekends and holidays when IT professionals are expected to perform all the disruptive work we do.

Tom Franken also helped review *Getting Started with XenDesktop® 7.x* by Packt Publishing.

Table of Contents

Preface

VMware PowerCLI is one of the most utilized command-line interfaces for a VMware vSphere System Administrator. Covering more than 480 different functions of a vSphere system, PowerCLI and PowerShell have become one of the staples of an automation enabler in this space. Taking the viewpoint of an administrator with some experience, *Essential VMware Administration with PowerCLI* introduces the idea and concept of taking the beginning steps toward developing one-line commands into multi-line scripts that can be used not only by the reader but also by others within their organization.

This book is designed with the mindset of think first, design second, script next, and test last. It covers getting the tool and integrating it with other products in the VMware stack and attempts to build on the knowledge outlined from the chapter before.

What this book covers

Chapter 1, An Introduction to Essential Administration with PowerCLI, sets the stage through discussing how to get PowerCLI, what the difference is between PowerShell and PowerCLI, discusses its version history, and provides a starting point with its installation. This chapter is a means to get you up and running with the right version and the best tools for the job.

Chapter 2, Comparing Point and Click Administration to PowerCLI and Scripting, looks through the eye of an administrator, helping to redefine the typical point and click doldrums and compare them to the exciting and provocative world of scripting.

Chapter 3, Enhancing the Scripting Experience, takes the previous chapter and builds upon it. This chapter will enhance the experience through better practices, help a team build a repository, and make the code reusable.

Chapter 4, Windows Administration within VMware Administration, starts with the preparation of a DevOps practice and the roles around it. This programmatically helps bridge the gap for becoming operationally transformed and influences how an organization can build a private cloud type of environment. We will use this mindset to build and provision a Windows script host, and run PowerCLI and Windows-based PowerShell in the same script.

Chapter 5, Workflows and vRealize Orchestrator, introduces the vRealize Orchestrator product, how workflows are developed, and where to use them. The chapter will provide a walkthrough of the installation of vRealize Orchestrator and where it fits in the environment.

Chapter 6, Running Workflows with Other VMware Products, discusses other VMware products such as NSX, Orchestrator, vRealize Operations Manager, Site Recovery Manager, and VSAN. It takes each technology and product, explains the product, where it fits, and then, finally, how PowerCLI can integrate with them.

What you need for this book

This book depends upon you, the reader, to have a vSphere environment to connect to and run the scripts against. This can be achieved through a try and buy download from VMware.com or with a company Test environment. A vSphere environment typically includes:

- A VMware vCenter server utilizing a Windows-based standalone installation or a vCenter Server Appliance
- An attached VMware ESXi server

This book was written using PowerShell v3 on Windows Server 2008 R2 VM or on PowerShell v4 running on Windows Server 2012 R2 VM. The scripts were tested on multiple systems for compatibility and to reduce errata in the scripts. PowerCLI version 6.0 R1 and R2 were used.

Who this book is for

PowerCLI Essentials is focused toward virtualization professionals and system administrators who want to discover and learn about the automation techniques associated with PowerCLI for complex virtual environments.

Conventions

In this book, you will find a number of text styles that distinguish between different kinds of information. Here are some examples of these styles and an explanation of their meaning.

Code words in text, database table names, folder names, filenames, file extensions, pathnames, dummy URLs, user input, and Twitter handles are shown as follows: "The `Get-PowerCLIConfiguration` command will show three scopes of the configuration, `Session`, `User`, and `AllUsers`."

Any command-line input or output is written as follows:

```
$PSVersionTable.PSVersion.Major
```

New terms and **important words** are shown in bold. Words that you see on the screen, for example, in menus or dialog boxes, appear in the text like this: "Set **Screen Buffer Size** for **Width** and **Height** whenever possible."

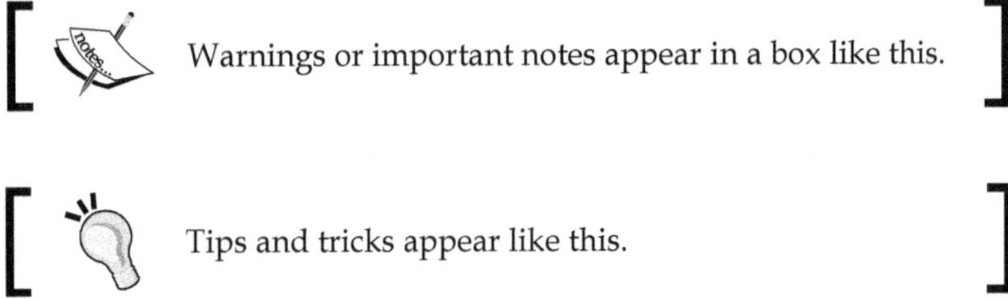

Warnings or important notes appear in a box like this.

Tips and tricks appear like this.

Reader feedback

Feedback from our readers is always welcome. Let us know what you think about this book—what you liked or disliked. Reader feedback is important for us as it helps us develop titles that you will really get the most out of.

To send us general feedback, simply e-mail `feedback@packtpub.com`, and mention the book's title in the subject of your message.

If there is a topic that you have expertise in and you are interested in either writing or contributing to a book, see our author guide at `www.packtpub.com/authors`.

Customer support

Now that you are the proud owner of a Packt book, we have a number of things to help you to get the most from your purchase.

Downloading the example code

You can download the example code files from your account at `http://www.packtpub.com` for all the Packt Publishing books you have purchased. If you purchased this book elsewhere, you can visit `http://www.packtpub.com/support` and register to have the files e-mailed directly to you.

You can download the code files by following these steps:

1. Log in or register to our website using your e-mail address and password.
2. Hover the mouse pointer on the **SUPPORT** tab at the top.
3. Click on **Code Downloads & Errata**.
4. Enter the name of the book in the **Search** box.
5. Select the book for which you're looking to download the code files.
6. Choose from the drop-down menu where you purchased this book from.
7. Click on **Code Download**.

Once the file is downloaded, please make sure that you unzip or extract the folder using the latest version of:

- WinRAR / 7-Zip for Windows
- Zipeg / iZip / UnRarX for Mac
- 7-Zip / PeaZip for Linux

Errata

Although we have taken every care to ensure the accuracy of our content, mistakes do happen. If you find a mistake in one of our books—maybe a mistake in the text or the code—we would be grateful if you could report this to us. By doing so, you can save other readers from frustration and help us improve subsequent versions of this book. If you find any errata, please report them by visiting `http://www.packtpub.com/submit-errata`, selecting your book, clicking on the **Errata Submission Form** link, and entering the details of your errata. Once your errata are verified, your submission will be accepted and the errata will be uploaded to our website or added to any list of existing errata under the Errata section of that title.

To view the previously submitted errata, go to `https://www.packtpub.com/books/content/support` and enter the name of the book in the search field. The required information will appear under the **Errata** section.

Piracy

Piracy of copyrighted material on the Internet is an ongoing problem across all media. At Packt, we take the protection of our copyright and licenses very seriously. If you come across any illegal copies of our works in any form on the Internet, please provide us with the location address or website name immediately so that we can pursue a remedy.

Please contact us at `copyright@packtpub.com` with a link to the suspected pirated material.

We appreciate your help in protecting our authors and our ability to bring you valuable content.

Questions

If you have a problem with any aspect of this book, you can contact us at `questions@packtpub.com`, and we will do our best to address the problem.

1

An Introduction to Essential Administration with PowerCLI

It's 4 pm on a Friday afternoon; you are packing up for a weekend camping trip with the family when your boss walks up to your desk with that *I have an immediate task that I need to hand off before I leave for the weekend* look on his face. You say, "Hey sir, I am just heading out camping this weekend and I need to beat the traffic, gotta go!" But you are just not that lucky, he quickly replies that his boss has an important meeting and needs some numbers before a Monday morning meeting, and adds, "I don't have the skills in the infrastructure to get this information and I am desperate to get this tonight. Besides, we will all look like heroes if we do this." "Or at least you will", you mutter under your breath.

Is that a normal scenario in your office? It was my role for a number of clients over the last few years until I started predicting the future and writing PowerCLI scripts. This book is a collection of administration experiences for managing a virtual infrastructure and for incorporating the vast skill of DevOps and scripting.

This chapter deals with the getting started mentality, experiencing taking the System Administrator role to the next level, and making the endless job of system operations a bite-sized effort.

In this chapter, you will learn about:

- PowerShell versions
- PowerCLI versions
- Getting PowerShell and PowerCLI
- Setting up the environment
- PowerShell basics

Why this book, and why now?

System Administration has always been somewhat of a thankless job, and as an admin, making things easier and quicker has always been the end goal. Understanding the underpinnings of PowerShell and PowerCLI will help with mundane daily tasks associated with regular administration. This chapter will help you get up to speed using the command line and allow shortening of time for you to be comfortable and proficient with it.

Understanding PowerShell versions

PowerShell, or **PoSH** as it is also known, is a shell scripting language that Microsoft created to replace batch and VBScript:

- Version 1 was introduced in 2006, was available to be downloaded for Windows XP, Vista, or Windows Server 2003, and was an optional component of Windows Server 2008
- Version 2 was integrated into Windows 7 and Windows Server 2008 R2 server and was available for download for earlier versions Windows XP SP3, Vista SP1, Server 2003 SP2, and Server 2008
- Version 3 was included in the base version of Windows 8 and Windows Server 2012 and could be installed on Windows 7 SP1 and 2008 R2 SP1
- Version 4 was integrated into Windows 8.1 and Windows Server 2012 R2 and could be installed on Windows 7 SP1, Windows Server 2008 R2 SP1, and Windows Server 2012
- Version 5 was still public beta as of the time of writing this book and is only installable on Windows 8.1, Server 2012, and 2012 R2

The basic structure of PowerShell

PoSH, based on the .NET framework, establishes a structure for programming/scripting into a human-readable format. Previous Windows scripting languages were, for the most part, cryptic and it wasn't always easy to pass a script to another administrator without the other administrator having extensive experience with the language.

Originally, the **Command Line Interpreter** or **CLI** (`cmd.exe` or `command.com`) allowed the running of specific executables or EXEs through a recipe of other known commands into a batch (`.bat`) or command script (`.cmd`). These scripts were used to build a setup for users (such as login scripts) or to launch a customized experience for an application startup or application installation.

The `cscript.exe` command enabled the inclusion of JavaScript or VBScript in the recipe and gave an administrator much more power to do far more complex and useful tools. The script was written with an external program (`notepad.exe`, for example) and then the `cscript.exe` command would have to precede the script.

After Microsoft's previous attempts at power and user friendliness, PoSH follows a basic human readable format of `Verb-Noun` for commands to be written. `Get-Service`, for example, uses the verb `Get` and the noun `Service` to specify the desire to get what services are running. Switches such as `DisplayName` can add to the clarity of the command and shorten the output:

```
Get-Service -DisplayName "VMware*"

Status     Name                 DisplayName
------     ----                 -----------
Running    VMTools              VMware Tools
Running    VMUSBArbService      VMware USB Arbitration Service
Stopped    vmvss                VMware Snapshot Provider
Running    vmware-view-usbd     VMware View USB
Running    wsnm                 VMware View Client
```

Although, not all commands contain switches such as get-service, clarification of this topic will be discussed later in this chapter.

Why is version understanding important?

As a typical System Administrator of multiple systems, it is likely that PoSH has been used due to newer Microsoft systems' (Exchange, SQL, or basic Windows admin functions) requirements for administration. With every version increment, additional features get added to the product, such as syntax simplification and additional cmdlets (pronounced command-let); this provides more functionality and script simplification possibilities to ensure certain version baselines in the environment allow single source scripts and common programming specifications.

To simplify, using the newest version available allows more functionality and ease of use. To use an analogy of buying a car, getting a newer car generally provides more features, performance, and gadgets. The manufacturer learns from the shortcomings of previous years and makes a better product.

Many companies use a plethora of Windows servers and Windows desktops to support their business's attempt at maintaining a standard platform for operating, deployment, and implementation to reduce the amount of work. In most environments, there are a variety of server versions in place, from Windows Server 2003 to Windows Server 2012 R2, all supporting different applications and being supported by hundreds of vendors. Each iteration of the Server product may, and typically does, have a variety of components that need to be maintained, such as Windows Patches, and in the case of VMware Infrastructure, VMware Tools and Hardware versions (Drivers).

As seen in the *Understanding the PowerShell versions* section, there are five listed versions available. Of these, **version 2** (**v2**) and **version 3** (**v3**) apply the greatest variance amongst all of the versions. Programming scripts using the v3 syntax will produce errors in v2, whereas v2 syntax will work in v3. Microsoft provides a much simpler experience using the newer versions, and all scripts within this book will be written using v3 unless otherwise stated.

Thankfully, there is a simple way to test what version of the Windows OS is running. To find out the current version of PoSH, type this command:

```
$PSVersionTable.PSVersion.Major
```

It will produce a very brief output, as follows:

```
3
```

v2 or v3 – what's the difference?

Is there a difference? There are quite a few variations between the two, but a single key differentiator shown here highlights it.

A basic one-line v2 command is as follows:

```
Get-VM | where {$_.Name -eq "Server"}
```

The same command using the v3 syntax is as follows:

```
Get-VM | where Name -eq Server
```

At first glance, the differences are apparent. v2 demands additional brackets, quotes and variable definitions to make the command work, whereas, v3 removes the cryptic demands and simplifies the command string. With v3, it is backwards compatible, so PoSH v2 commands will also work within the v3 command structure.

One of the significant additions to v3 is the inclusion of workflows. Workflows allow the running of multiple scripts in a cascading fashion. This provides a means to develop the ability to write recipes with many parts into a single consistent plan. This can be extremely powerful and useful and is discussed in later chapters.

So the question that begs to be asked is: *Why not develop all code within the context of v2?* As seen earlier, PoSH v3 simplifies the development of the code base, so why not use this? This is the dilemma of every site or company embarking on a path to automation, whether to choose a path of conformity versus the added administration of functionality. As an Administrator, most environments that are worked on will include a variety of different systems, from 2003 Server, 2008 R2 Server, and 2012 Server, which include different versions of PowerShell. One of the first automation scripts that will be shown will be a sample bit of code to look through the environment and find PowerShell versions and OSs with the installed version of PowerShell.

Because this book deals with workflows and automation in the latter chapters, all code will be written in v3. **Version 4 (v4)** provides many additional cmdlets but doesn't provide a significant interface change to warrant the inclusion of it. If there is a v4 specific command used, the text will ensure the reader's understanding of such.

Installing PowerShell v3 on a Windows 7 or Windows 2008 R2 machine

On a standard Windows 7 computer, PoSH v2 comes standard as referred to earlier in this chapter. If running the PoSH command, as shown in the last section, reveals that the system is indeed running v2, there is a simple process that can be gathered from this Microsoft URL:

`http://www.microsoft.com/en-us/download/details.aspx?id=34595`

File Name	Size
Windows6.1-KB2506143-x64.msu	15.8 MB
Windows6.0-KB2506146-x64.msu	14.4 MB
Windows6.0-KB2506146-x86.msu	10.5 MB
Windows6.1-KB2506143-x86.msu	11.7 MB
WMF 3 Release Notes.docx	53 KB

Listing of available files

The one point it doesn't list is the fact that because PoSH v3 uses the .NET framework for the majority of its functions, .NET version 4 at minimum must be installed on the system as well. Both components can be pushed through any patch management framework that may be in use.

Understanding PowerCLI

VMware created an **Application Program Interface** (**API**) that provides a way for third-party vendors and developers to have other software products register and control the vSphere environment. PowerCLI is a group of functions, written in PowerShell, that convert the cryptic commands of the API into a human-readable format to simplify the scriptwriter's job. For example, one of the first commands you write is `Connect-VIServer`. This single command is programmed as a group of API calls to initiate a connection to the vCenter server, pass user credentials, and open a network socket to the server. Understanding all of the pieces of how to connect to a vCenter server is now irrelevant and the single command is used.

Each edition of vSphere brought with it a long list of features and options permitting the user to do more and more with the Virtual Environment, and with that, more and more cmdlets needed to be written for PowerCLI to enable better control of the environment. One of the best improvements was in vSphere version 4 and was the creation of **Distributed Virtual Switches** (**DVS**). PowerCLI had no means to create, manage, or delete such switches, and on numerous occasions, the API had to be used to perform the function. So, with each additional PowerCLI version, the developers of the toolset brought it closer and closer to feature parity. As of version 6 release 1 (R1), the two products are closer than they have ever been. So, having the latest PowerCLI version is optimal for administering the environment as the newest PowerCLI version is still very capable of controlling older versions of vSphere.

Getting PowerCLI versions

To get the PowerCLI version, either watch as you open the interface or type this command:

```
$(Get-PowerCLIVersion).Major
```

The preceding command line will produce an output as follows:

```
6
```

Otherwise, you can type:

```
$(Get-PowerCLIVersion).UserFriendlyVersion
```

Now, this command will produce an output as follows:

```
VMware vSphere PowerCLI 6.0 Release 1 build 2548067
```

As shown, the output can be simple or human readable. Either way works for general examination of the command, but both can be used for different purposes. For example, if there is a need to run a specific command that is only available in version 6, then a query of this command may be ideal to stop the display of an error.

The PowerCLI change log

These are the highlights of PowerCLI right from version 1.0.1 (when it was named VI Toolkit) up to the latest version 6.0:

- **VI Toolkit 1.0.1 (released in September 2008)**: The VI Toolkit was a utility that provided the administrator with simple commands to gather and check the environment for simple information. Even though the majority of the tools still commanded a grasp of the API, the commands provided a huge advantage for conducting repeatable tasks.
- **VI Toolkit 1.5 (released in January 2009)**: The release of 1.5 added quite a few new cmdlets, fixed a few bugs in the older version, and notably added `Set Verb` to the cmdlets. The key differentiator was now being able to do more than read and report.

  This version also sparked a number of key players in the community such as Virt-Al, LucD, and ESloof to post numerous blog posts and tons of VMware Community posts on how to do stuff with PowerShell and VMware. This timeframe also sparked the creation of the Virtualization EcoShell Interface and caused the PowerGUI communities to pop up, adding much more credibility to the line.

- **PowerCLI 4.0 (released in May 2009)**: Here is the first of the PowerCLI products that aligned with the vSphere release. The highlights were around the Host Profiles and Host security and introduced the `Invoke-vmscript` function.
- **PowerCLI 4.0.1 (released on November 19, 2009)**: As this was a minor release, it added a few more cmdlets and fixed some bugs from the previous build. It started the OS customization portion of the command set, and brought about Power state controls for the Hosts.
- **PowerCLI 4.1 (released on July 13, 2010)**: This version started the tradition of being released shortly before VMworld to help coincide with the release of the vSphere version. There were very few new cmdlets, but numerous enhancements and bug fixes for previous versions.

- **PowerCLI 4.1 U1 (released on December 1, 2010)**: The largest addition to this release was the ability to run ESXCLI (usually reserved for the command line interface on the host, the vSphere CLI command set, or the VMware Management Appliance). This added a huge subset of commands to the Administrator's tool chest with regard to scripts.

  Another feature was the ability to control Distributed vSwitches, through the scripts. This allowed the ability to build an environment through a script by initiating the host profiles, as listed earlier, and now build a distributed vSwitch and provision VMs network interface cards to said dvSwitch.

- **PowerCLI 5.0 (released in August 24, 2011)**: The 5.0 release introduced feature parity to the Storage I/O control set and was available for download at the same time as the 5.0 vSphere release. Most of this release added functionality to what was already there and fixed a few bugs in the `Invoke-VMScript` cmdlet.
- **PowerCLI 5.0.1 (released on January 9, 2012)**: This release added vCloud support to the stack.
- **PowerCLI 5.1 R1 (released on September 10, 2012)**: Shortly after VMworld, when vSphere 5.1 was released, v 5.1 was available. This release brought support for **Storage Distributed Resource Scheduling** (**SDRS**), Storage vMotion, and added numerous enhancements to the vCloud Director product.

  This version also brought a new modification to the tool; it deprecated commands for the first time in the product line. This was significant because a script using the command `Set-VMHostAdvancedConfiguration` would have an error because the command had changed.

- **PowerCLI 5.1 R2 (released on February 11, 2013)**: Further dvSwitch networking support was added to the cmdlet list, including the ability to export the configuration for product migration. There were a number of vCloud enhancements and additional deprecation of cmdlets.
- **PowerCLI 5.5 R1 (released on September 19, 2013)**: Version 5.5 brought a few more dvSwitch cmdlets, introduced tags, allowed management of VSAN, and introduced additional component pieces to the install.
- **PowerCLI 5.5 R2 (released on March 11, 2014)**: Release 2 of the 5.5 version added support for PowerShell v4.0, enhanced the tag cmdlets, and at long last, provided 64-bit capability for OS customization. This version also exposed an ability to communicate with a **Site Recovery Manager** (**SRM**).

- **PowerCLI 5.8 R1 (released on September 9, 2014)**: 2014 was the first VMworld that didn't release a new vSphere version. The enhancement that year was in the upgrade to the suite of products. There was an incremental version numbering of PowerCLI, SRM, vCenter Operations Manager, Infrastructure Navigator, and other related products to version 5.8. The list of enhancements for this version of PowerCLI was also one of the smallest, but the tool had its startup time increased and an enhancement to the error reporting within the tool.
- **PowerCLI 6.0 (released on March 12, 2015)**: This release coincided with the release of vSphere 6 and provided support for VVOLs and VSAN enhancements and moved some of the cmdlets to be supported as modules instead of snap-ins, as previously performed. The change log discusses re-examining the script code to support this change. It is a fundamental change to the PowerCLI product line, and the scripts in this book will be reflective of this version.

  The version listing, at first glance, may seem irrelevant, but as the product has changed and grown, the script methodology and syntax has changed. From personal experience in numerous client sites, there is a mixture of different versions installed at different times by different administrators. This inconsistency, like the PoSH v2 and v3 section previously discussed, can and will cause scripting errors within the automation of the site.

> Further reading of the changes can be done at:
> `https://www.vmware.com/support/developer/PowerCLI/changelog.html`

Getting PowerCLI

To get PowerCLI, there are a few places to start. Google always brings the browser first to the VMware blog site, which is all right as that usually has the correct download link. However, the link is provided later for all the versions of PowerCLI.

> It is always better to download direct from VMware. There are other links that can be used to download from, but because it is a programmable interface directly within a production virtual environment, it is ideal to get it from the source.

Pre-requisites before getting PowerCLI

Having a VMware ID will be required to be able to download PowerCLI.

Downloading PowerCLI

To download the latest version, perform the following steps:

1. Go to `https://www.vmware.com/support/developer/PowerCLI/`, and you will see the following window:

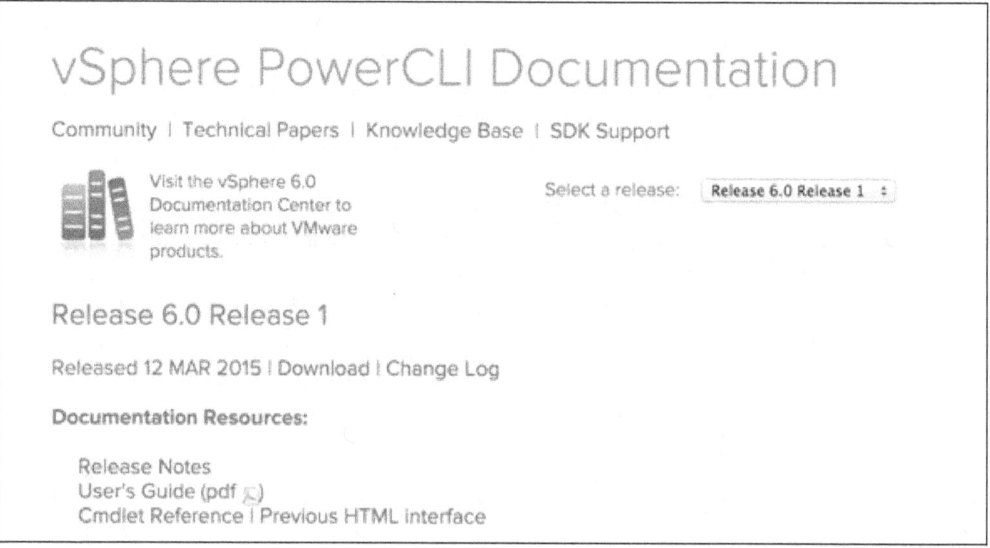

2. Click on the download link shown in the screenshot and the web page takes the browser to the login page.
3. Enter the VMware ID and the password, and the PowerCLI executable will be downloaded.

Installing PowerCLI

The typical installation of PowerCLI, whether it is 4.0 or 6.0, includes the VIX API (discussed earlier), and newer versions include the Remote Console as well. The remote console provides a remote console to any available VM running in the virtual infrastructure.

The installation shown in the following diagram is PowerCLI Version 6.0 on a Windows Server 2008 R2 platform. Let's follow the installation process:

1. When launching the installation, the shown interface seems a little strange to typical installs:

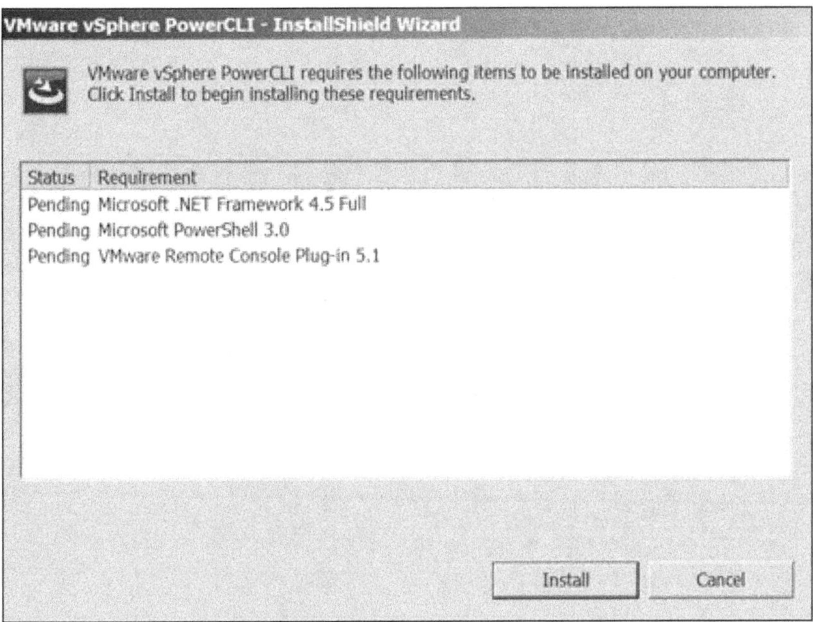

2. However, the installer runs separate installations of the components, one for the required **.NET Framework 4.5 Full**, and the PowerShell v3 and remote Console Plug-in 5.1 after. First, the installer downloads the file as follows:

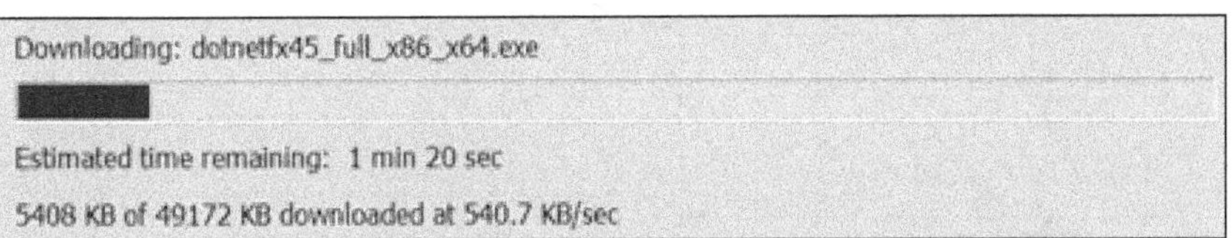

3. Once the file is downloaded, the installer begins the installation process:

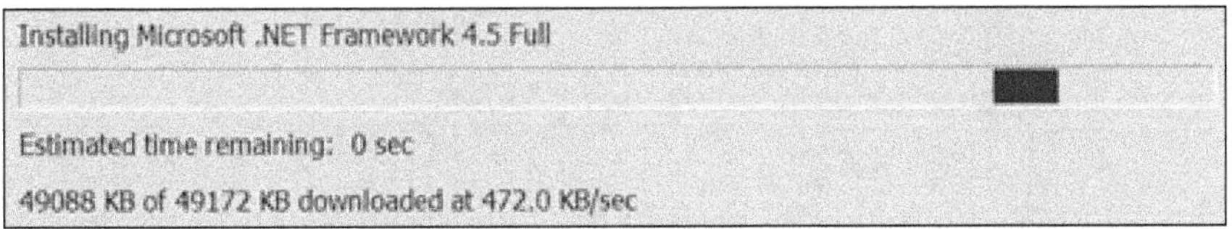

4. Then, the installer runs the installation file:

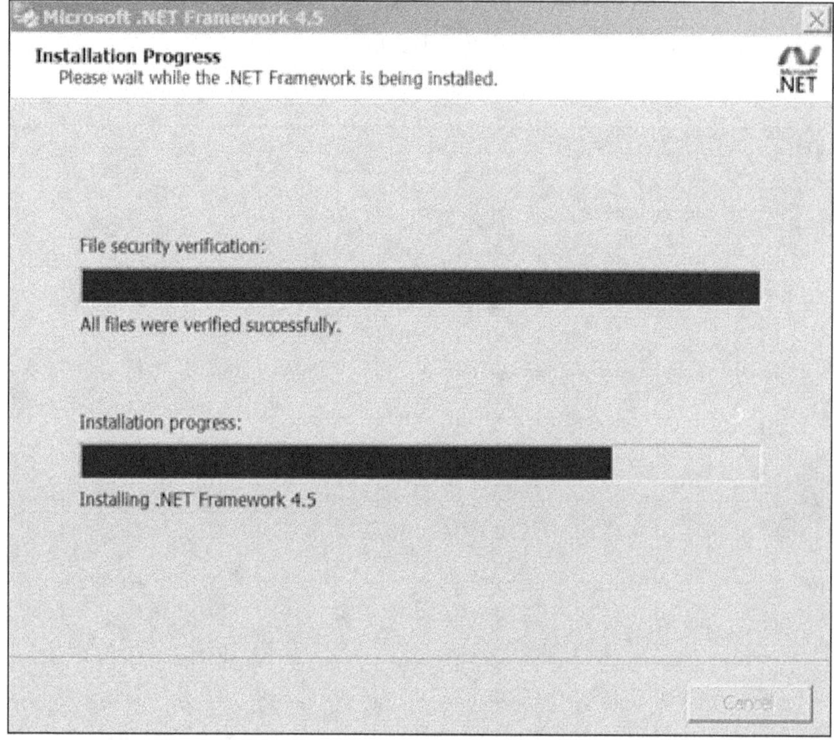

Once the .NET framework is installed, PowerShell v3 is installed too.

Don't be surprised if the installation fails at this point. When running through this procedure, the following error popped up:

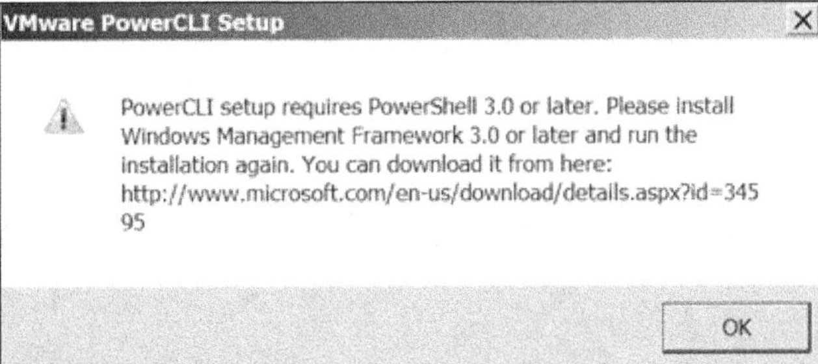

This popup demands that the Windows Management Framework version 3.0 be installed. (Installing WMF 3.0 is shown in the *Installing PowerShell v3* section earlier in the chapter.) The installation will exit with an error if this isn't done. Allow the installer to exit, install WMF 3.0, and rerun the installer.

5. Once complete, the VMware Remote Console Plug-in is installed. The remote console displays where it will install the web browser plugins on the system.

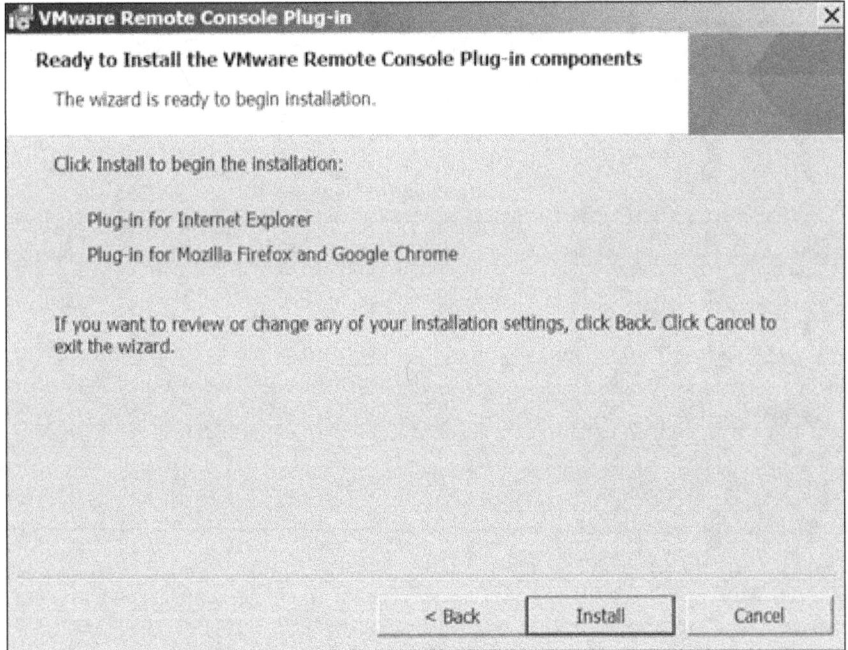

6. The Remote Console is installed and the PowerCLI snap-in component is started. A warning about setting the `RemoteSigned` execution policy is displayed, as shown in the following screenshot (this will be further explained in the next section):

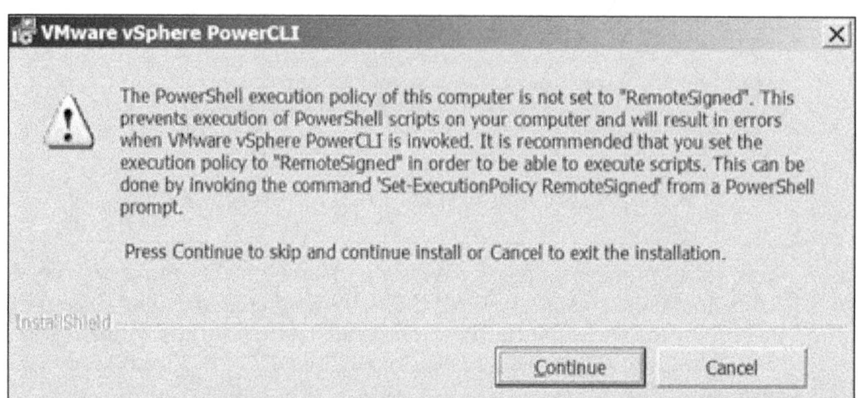

7. Then, finally, the PowerCLI installation begins after clicking on **Continue**, which brings you the Welcome screen. Click on **Next** to continue to the installation wizard for PowerCLI:

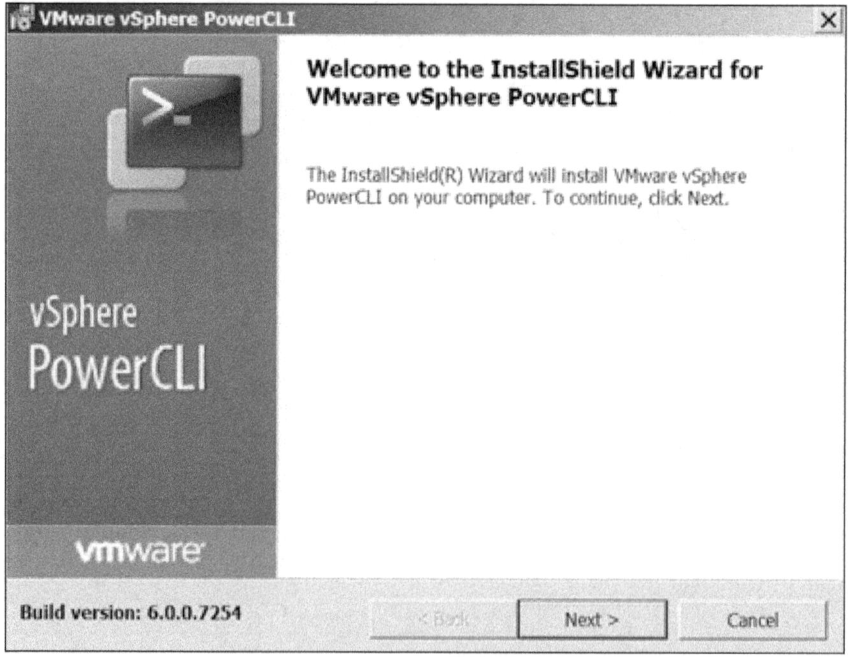

8. This will bring you to the **VMware Patents** window. Click on **Next** to continue. This will bring you to the **License Agreement** window.
9. Accept the License Agreement, choose the components, set the path of installation, and run the installation process. Finally, you will reach the setup process as shown in the following screenshot:

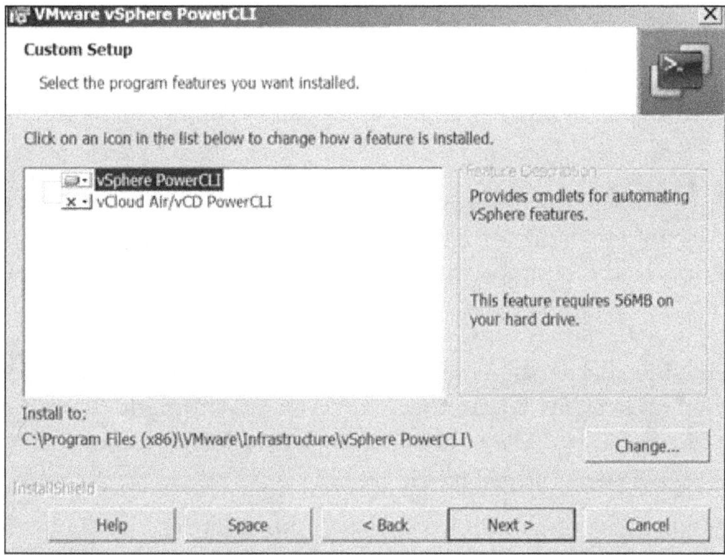

On older versions, the VIX API is shown in the installer, but is not shown in the newer versions.

The final list of applications from a base Windows 2008 R2 server installation looks like this:

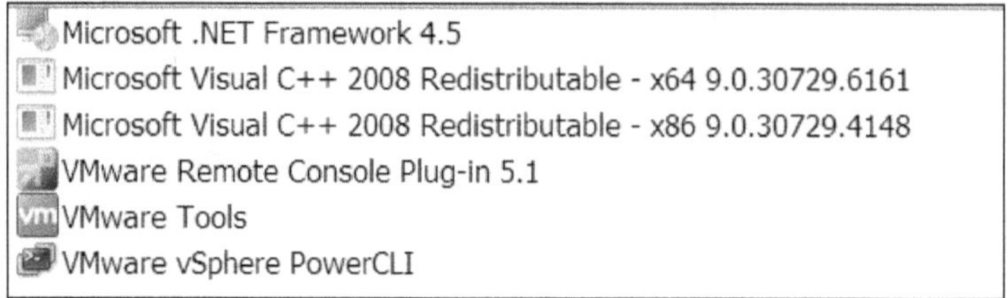

Getting help

PoSH v2 installs natively and includes all the help files as installed on a system. Version 3 changes that dichotomy in to forcing the user to download or access the help online. On PoSH v3, the Update-Help command should be run to ensure all help files are downloaded on to the system where the scripts will to be written. This helps reduce the file space usage on systems not being used for script development. If help is needed on systems where the Update-Help command has *not* been run, a Get-Help <command> -online command can open an associated web page with the appropriate command syntax shown.

Once the help files are downloaded, a simple query such as Get-Help Get-VM will produce a limited explanation of the command with the name of the command, synopsis, syntax, description, related links (similar commands), and remarks. The remarks section will remain the same for every help request, as it outlines alternate switches in the getting help section.

The Get-Help command has an alias of Help that can be also be used for short hand. For example, Get-Help Get-VM can be written as Help Get-VM.

For most help requests, the scriptwriter either needs an understanding of the syntax or to see examples of the command in use. Using the -examples switch displays the example code built into the help files directly. So typing Get-Help Get-VM -examples displays the following output:

```
NAME
    Get-VM


SYNOPSIS
```

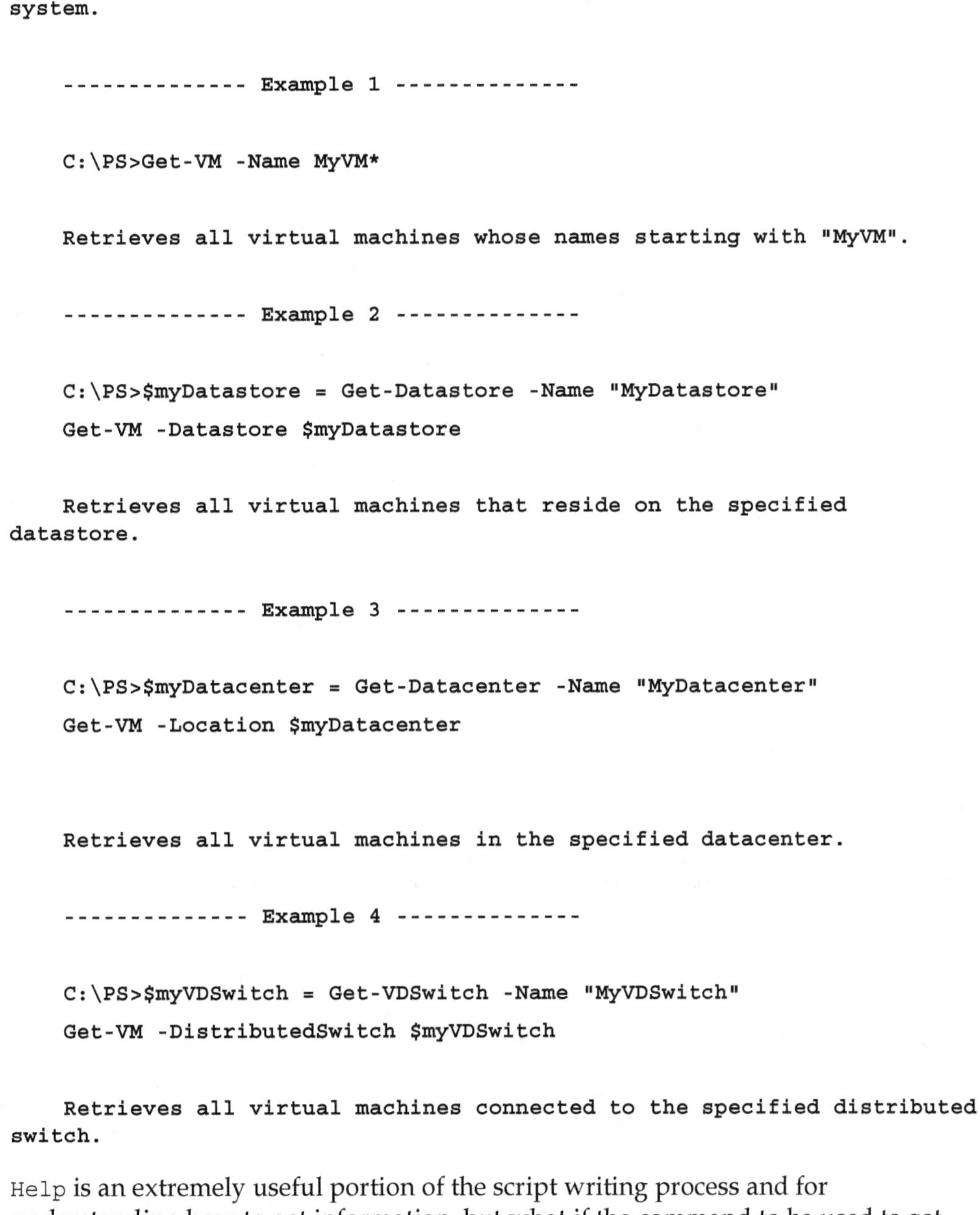

```
    This cmdlet retrieves the virtual machines on a vCenter Server
system.

    -------------- Example 1 --------------

    C:\PS>Get-VM -Name MyVM*

    Retrieves all virtual machines whose names starting with "MyVM".

    -------------- Example 2 --------------

    C:\PS>$myDatastore = Get-Datastore -Name "MyDatastore"
    Get-VM -Datastore $myDatastore

    Retrieves all virtual machines that reside on the specified
datastore.

    -------------- Example 3 --------------

    C:\PS>$myDatacenter = Get-Datacenter -Name "MyDatacenter"
    Get-VM -Location $myDatacenter


    Retrieves all virtual machines in the specified datacenter.

    -------------- Example 4 --------------

    C:\PS>$myVDSwitch = Get-VDSwitch -Name "MyVDSwitch"
    Get-VM -DistributedSwitch $myVDSwitch

    Retrieves all virtual machines connected to the specified distributed
switch.
```

`Help` is an extremely useful portion of the script writing process and for understanding how to get information, but what if the command to be used to get, say, VM information is not known? Running the `Get-Help VM` command will show every cmdlet available with VM in the name with a brief synopsis of each one.

Another option is to use a search engine to locate an example of the certain task or information needed. For example, searching for "PowerShell get VM guest information" produces a number of results from the `Get-VMGuest` command to utilizing `Get-View`, and even discusses some methods to get this information from a Hyper-V server. Where too much information is daunting, understanding the more specific information to be searched for will mean the more relevant information should be returned.

Lastly, never be afraid to ask. VMware communities were built on this premise and some of the best scripters asked questions at one point.

Setting up the PowerCLI installation

Once the installation is complete, a couple of minor tweaks are both required and recommended. The first required portion of the environment setup is to set the execution policy. The execution policy definition is taken directly from the PoSH help file:

```
Get-Help About_Execution_Policies


Windows PowerShell execution policies let you determine the conditions
under which Windows PowerShell loads configuration files and runs
scripts.


You can set an execution policy for the local computer, for the current
user, or for a particular session. You can also use a Group Policy
setting to set execution policy for computers and users.


Execution policies for the local computer and current user are stored in
the registry. You do not need to set execution policies in your Windows
PowerShell profile. The execution policy for a particular session is
stored only in memory and is lost when the session is closed.


The execution policy is not a security system that restricts user
actions. For example, users can easily circumvent a policy by typing
the script contents at the command line when they cannot run a script.
Instead, the execution policy helps users to set basic rules and prevents
them from violating them unintentionally.
```

Of all the Execution Policies, `Restricted` is enabled by default:

```
"Restricted" is the default policy


  Restricted
```

```
- Default execution policy.
- Permits individual commands, but will not run scripts.
- Prevents running of all script files, including
  formatting and configuration files (ps1xml), module
  script files (.psm1), and Windows PowerShell
  profiles (.ps1).
```

After reading the `About_Execution_Policies` file about execution policies and reading this default policy, it is clear to see that this will not allow the interface to be very useful when running PowerCLI. In fact, if the PowerCLI icon is clicked on without setting the execution policy to a less restrictive setting, the snap-in will not load.

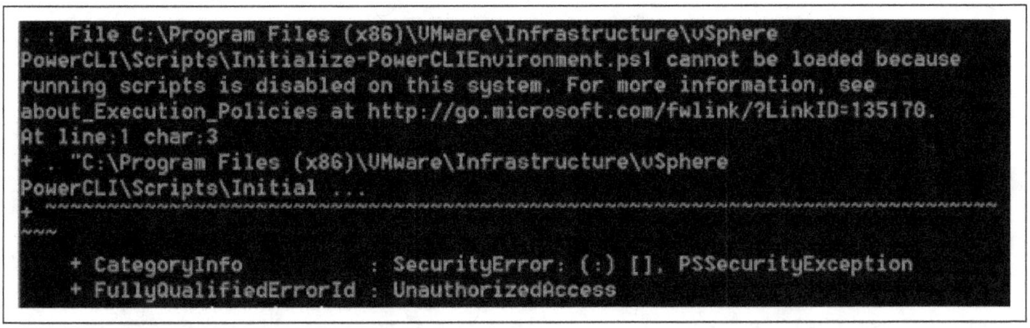

```
. : File C:\Program Files (x86)\VMware\Infrastructure\vSphere
PowerCLI\Scripts\Initialize-PowerCLIEnvironment.ps1 cannot be loaded because
running scripts is disabled on this system. For more information, see
about_Execution_Policies at http://go.microsoft.com/fwlink/?LinkID=135170.
At line:1 char:3
+ . "C:\Program Files (x86)\VMware\Infrastructure\vSphere
PowerCLI\Scripts\Initial ...
+ ~~~~~~~~~~~~~~~~~~~~~~~~~~~~~~~~~~~~~~~~~~~~~~~~~~~~~~~~~~~~~~~~~~~~~~~~~~~~~
~~~
    + CategoryInfo          : SecurityError: (:) [], PSSecurityException
    + FullyQualifiedErrorId : UnauthorizedAccess
```

It shows it in red to ensure that the error is read and understood. To bypass this, the `Set-ExecutionPolicy` command must be run under an administrator credential. Right-click on the icon for PowerCLI and select **Run as Administrator**. This enables the User Access Control for the PowerCLI/PowerShell window.

One of the following options must be selected:

```
AllSigned


- Scripts can run.
- Requires that all scripts and configuration files be signed be a
trusted publisher, including script that you write on the local computer.
- Prompts you before running scripts from publishers that you have not
yet classified as trusted or untrusted.
- Risks running signed, but malicious, scripts.


RemoteSigned


- Scripts can run.
```

```
- Requires a digital signature from a trusted publisher on script and
configuration files that are downloaded from the Internet (including
e-mail and instant messaging programs).

- Does not require digital signatures on scripts that you have written on
the local computer (not downloaded from the Internet).

- Runs Scripts that are downloaded from the Internet and not signed, if
the scripts are unblocked, such as by using the Unblock-File cmdlet.

- Risks running unsigned scripts from sources other than the Internet and
signed, but malicious, scripts.


UnRestricted


- Unsigned scripts can run. (This risks running malicious scripts.)

- Warns the user before running scripts and configuration files that are
downloaded from the Internet.


Bypass


- Nothing is blocked and there are no warnings or prompts.

- This execution policy is designed for configurations in which a
Windows PowerShell script is built in to a larger application or for
configurations in which Windows PowerShell is the foundation for a
program that has its own security model.


Undefined


- There is no execution policy set in the current scope.

- If the execution policy in all scopes is Undefined, the effective
execution policy is Restricted, which is the default execution policy.
```

Typically, most environments use the `RemoteSigned` option as this allows the functionality to run scripts on other computers and allows some additional security from unsigned scripts. If the environment being set up is a highly secure environment, the `AllSigned` option may be used when proper **Public Key Infrastructure** (**PKI**) is set up. This functionality will not be described in this book, and the `RemoteSigned` option will be assumed for the remainder of the book:

```
Set-ExecutionPolicy RemoteSigned
```

Once the `ExecutionPolicy` is set, configuration of the interface window should be next as it requires the additional permissions of the Administrator account. Right-click on the **Heading** window, and select **Properties**.

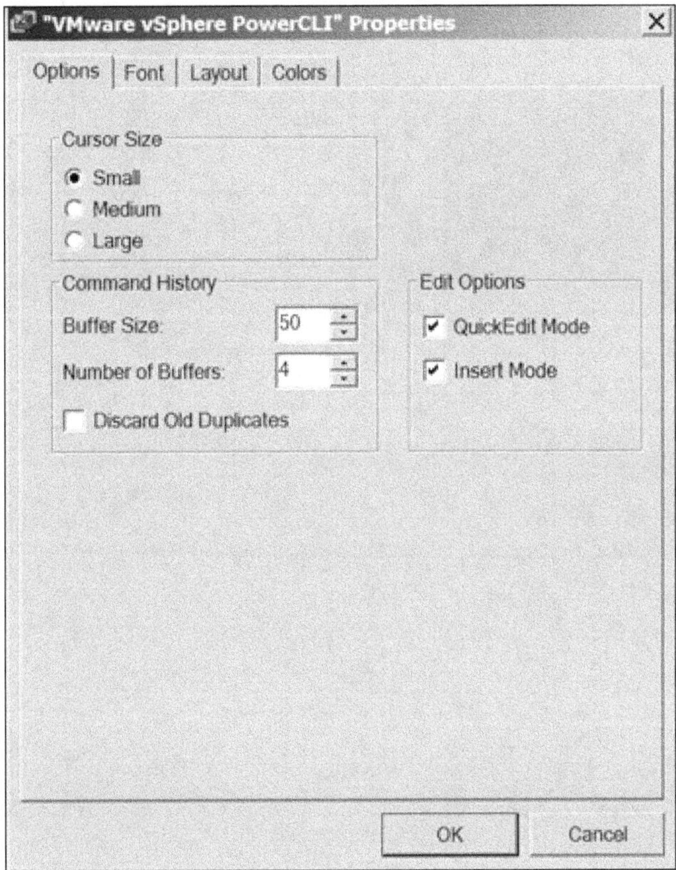

The `QuickEdit` mode allows the right-clicking, copying, and pasting functionality of the **Command** window. This feature helps tremendously when doing quick one-liners, needing report information or just wanting a portion of a command for another.

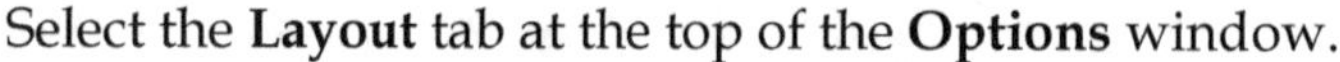

Select the **Layout** tab at the top of the **Options** window.

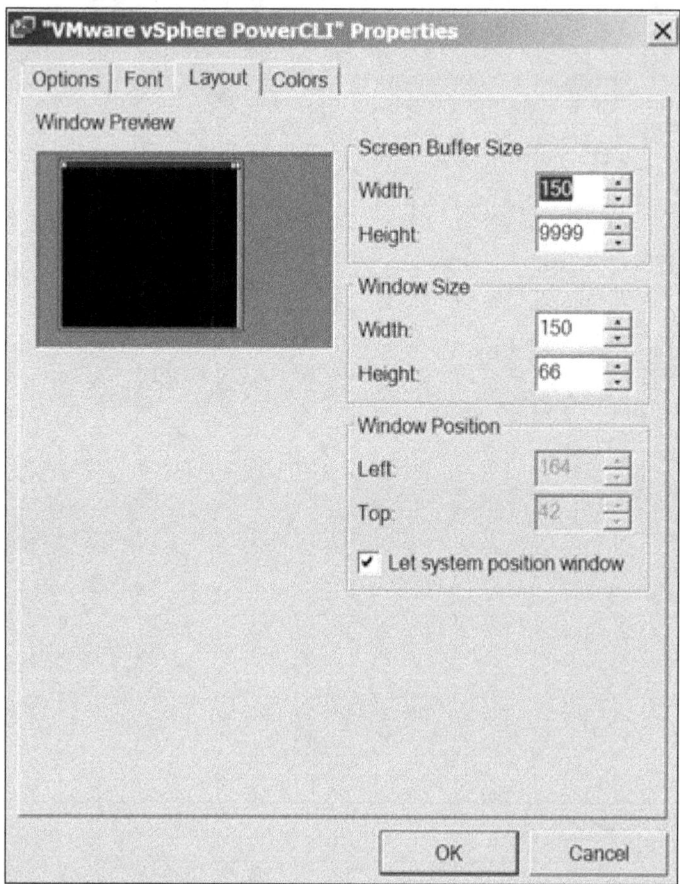

Set **Screen Buffer Size** for **Width** and **Height** whenever possible. The width buffer should match the actual width size. This permits the interface to need limited scrolling when getting output. The height buffer should be large, up to **9999**, but if memory is a concern for the window, reduce the buffer size. This buffer allows for the output of a command to exceed the window size. For example, if the `Get-VM` command produces a listing of 3000 VMs, having a small buffer may not allow the window to be scrolled back to view the entire output. If the buffer size is 250, then 250 lines of scroll back will be permitted.

The last thing to check is the PowerCLI configuration. If the environment being worked on has multiple vCenter servers, having the configuration support multiple systems simultaneously can be advantageous. By default, the configuration is to connect to a single vCenter at a time; however, if the session is connected to one vCenter and a second `Connect-VIServer` command is run, this message pops up on the interface window:

```
Working with multiple default servers?


Select [Y] if you want to work with more than one default servers. In
this case, every time when you connect to a different server using
Connect-VIServer, the new server connection is stored in an array
variable together with the previously connected servers. When you run a
cmdlet and the target servers cannot be determined from the specified
parameters, the cmdlet runs against all servers stored in the array
variable.
    Select [N] if you want to work with a single default server. In this
case, when you run a cmdlet and the target servers cannot be determined
from the specified parameters, the cmdlet runs against the last connected
server.


    WARNING: WORKING WITH MULTIPLE DEFAULT SERVERS WILL BE ENABLED
BY DEFAULT IN A FUTURE RELEASE. You can explicitly set your own
preference at any time by using the DefaultServerMode parameter of Set-
PowerCLIConfiguration.


[Y] Yes  [N] No  [S] Suspend  [?] Help (default is "Y"): y
```

The `Get-PowerCLIConfiguration` command will show three scopes of the configuration, `Session`, `User`, and `AllUsers`. The standard setup will show the following output:

```
Scope    ProxyPolicy     DefaultVIServerMode InvalidCertificateAction DisplayDeprecationWarnings WebOperationTimeout
                                                                                                  Seconds
-----    -----------     ------------------- ------------------------ -------------------------- -------------------
Session  UseSystemProxy  Multiple            Unset                    True                       300
User
AllUsers
```

Based on the preceding command connecting to the second vCenter, the configuration will show the following output:

```
Scope    ProxyPolicy     DefaultVIServerMode InvalidCertificateAction DisplayDeprecationWarnings WebOperationTimeout
                                                                                                  Seconds
-----    -----------     ------------------- ------------------------ -------------------------- -------------------
Session  UseSystemProxy  Multiple            Unset                    True                       300
User                     Multiple
AllUsers
```

Now, to allow `AllUsers` by default, the command to enable that would be as follows:

```
Set-PowerCLIConfiguration -DefaultVIServerMode Multiple -confirm:$False
```

Finally, the output would be displayed as follows:

```
Scope    ProxyPolicy     DefaultVIServerMode InvalidCertificateAction DisplayDeprecationWarnings WebOperationTimeout
                                                                                                 Seconds
-----    -----------     ------------------- ------------------------ -------------------------- -------------------
Session  UseSystemProxy  Multiple            Unset                    True                       300
User                     Multiple
AllUsers                 Multiple
```

Although there are many other environment options that can be performed, these, in the author's opinion, are the most useful.

Summary

In this chapter, we have explained what PowerShell is, what the basics of the versions are, and how to get it, install it, and get help for it. We also discussed how to implement VMware PowerCLI and how to setup the interface and provide consistency for the scriptwriter. Although, as the reader, most of this information likely is understood, it will establish a baseline to assist in future chapters with regard to versions and setup.

In the next chapter, we will deal with the adaptation of Point and Click administration to PowerShell commands, look at tools that can be used, and orchestrate a VM build script. We will examine how to do reporting, and change configuration on multiple VMs at the same time.

2

Comparing Point and Click Administration to PowerCLI and Scripting

It's now 5 pm on a Friday and the report script that your boss asked you to create is ready to run. As the prompts flash on the screen, you say "Yes! It is done". You quickly load your e-mail client and you send the script off. You quietly think, "I should include a fake invoice for $1 million dollars." and chuckle to yourself. You notice that the office is silent and empty, and you begin packing up to go when you notice the datastores beginning to go offline and the VMs going gray. Do you leave and hope the environment is okay or do you deal with this next crisis that is about to surface?

Daily administration can be simple and straightforward and turn to chaos in a matter of minutes. Scripting, automation, and centralized management and reporting are critical to any environment. Orchestration tools, centralized version stores, and prebuilding of commonly used scripts can be crucial when attempting to troubleshoot any potential problem. This chapter displays commonly used administration point and click tasks and converts them to either a single line PowerCLI script or a multiline, menu-driven interface. We'll be covering the following topics in this chapter:

- Discussing the freely downloadable tools
- Inclusion of vCheck health check
- Building your own scripts
- Single line power scripts
- Orchestrating a build script

Getting started

As an administrator, the desire is to produce reports on current system performance and configuration. Perhaps an audit of the drift of such systems is the typical use case that explains why PowerCLI has been downloaded in the first place. Most individuals start by downloading someone else's script and attempting to run it in the environment. This chapter deals with that—pinpointing safe and viable scripts to run within your environment.

Where do we begin?

One key tool that can be used to begin to understand the capabilities of scripting is **RVTools** (at `www.robware.net`), which provides a quick report of almost everything in the Virtual Environment. This truly is a fantastic example of what information can be garnished from a running vSphere/vCenter instance. It uses the proper **Software Developers Kit** (**SDK**) and the .NET framework to pull the information into a CSV file.

A PoSH-centric file named `vCheck`, originally started by the author of `www.virtual.net`, Alan Renouf, has blossomed into its own framework for environment reports and audits. There are a number of contributors to the project now and it has grown to encompass vChecks for vCloud Director, Microsoft Exchange, Microsoft System Center Virtual Machine Manager, and more.

These provide a base of knowledge if there is little to no experience or time to learn.

Where do we find scripts and snippets that are safe to run?

There are, literally, hundreds of sites, blogs, and developer sites that contain PoSH code and PowerCLI optimized scripts. This book is not going to list every single place to start; however, it is going to list key ways to validate the code and some places to start.

VMware communities

The VMware communities are a great place to start. Where not all code is good, there are key individuals that stand out as contributors of merit. LucD, Alan Renouf, RvdNieuwendijk, and others have built an impressive list of appropriate scripts for various tasks. Typical searches in a favorite search engine, `<Script purpose>` (site: `https://communities.vmware.com`), should find any number of examples of scripts to supply syntax and methods to run.

VMware blogs

Heading to `https://blogs.vmware.com/powercli` will lead you to many interesting articles about PowerCLI and different ways to program interesting areas of the utility. All the articles are written by VMware employees and focus on the VMware-centric ways to use the language.

VMware flings

Flings are the VMware engineer's tools and applications, which they developed outside of the typical workday, written to ease a personal inconvenience or just because they thought it would be cool. The URL is `https://labs.vmware.com/flings`, and it provides a plethora of PowerCLI components. One of the best tools is project Onyx, which provides the code based upon a certain task performed in the GUI. The project has been around for quite some time and now supports use through the WebClient.

Onyx is run as a man-in-the-middle component, where it intercepts the datastream between the Administrator's console and the vCenter; it reads the SOAP and API strings and parses it to a code string. The Administrator can copy and paste the code snippet and run it in a very similar way to how they ran it in the GUI. Some of the code bits are quite cryptic but if there is no associated cmdlet, and if the developers are in a jam, this tool will help.

PowerGUI

Previously an independently developed **Integrated Development Environment (IDE)**, PowerGUI, was the ideal free tool for the VMware administrator. The tool has gone through a number of acquisitions and has finally landed at Dell. It is still a great tool and provides many ways to help in the development of scripts (`http://en.community.dell.com/techcenter/powergui`).

Notepad++

This is another great and free IDE that provides color syntax, and other great features (`https://notepad-plus-plus.org/`).

PowerShell Scriptomatic

Microsoft has some excellent tools and utilities to help in the development cycle as well. This website (`https://technet.microsoft.com/en-us/library/Ff730935.aspx`) provides a tool from Ed Wilson that allows a script development for WMI calls. As an Administrator, WMI namespaces have proven to be a treasure trove of valuable configuration data from a Windows OS implementation.

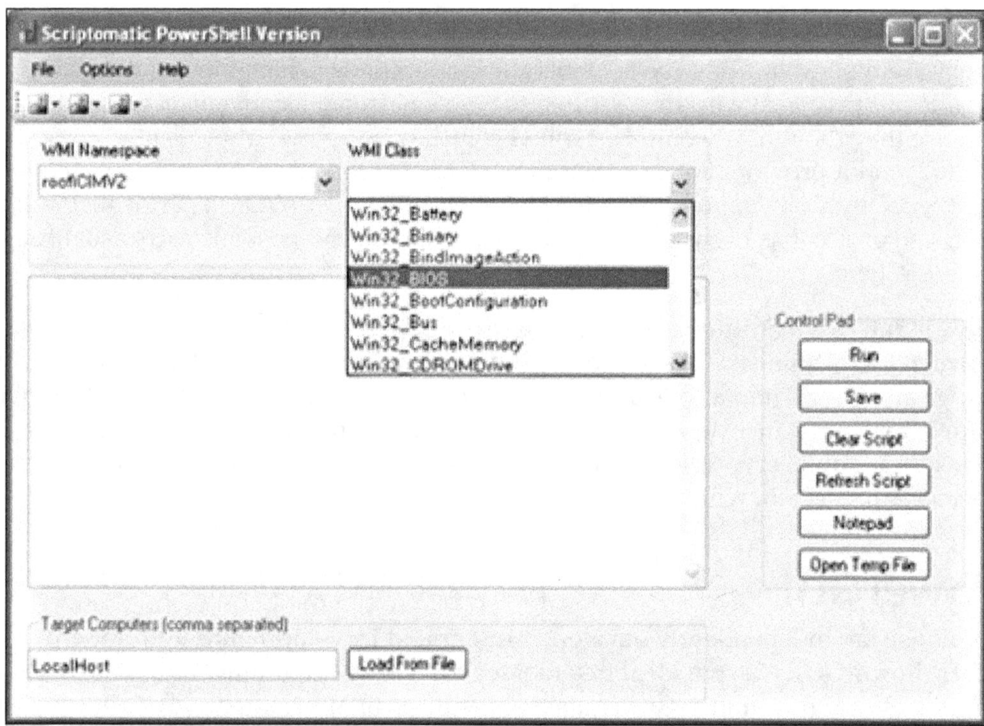

Microsoft TechNet Script Center

Script Center offers some fantastic tools, such as Script Browser and Script Analyzer, and numerous other script repositories for older script coding. This can be found at `https://technet.microsoft.com/en-us/scriptcenter`. These are a few examples of tools that can help you on the journey.

Programming interfaces

PowerShell, by default, includes the **Integrated Scripting Environment** or **ISE**. This application provides instant help reference, command completion, syntax prompting, an integrated PowerShell interface, and coloring for visual reference. It does not support PowerCLI out of the proverbial box and adding it is also not available at this point in the lifecycle of the product; however, starting each script with the following code snippet will ensure PowerCLI commands are included in the running of the script.

```
Add-PSSnapinVMware.VimAutomation.Core -ErrorAction SilentlyContinue
```

This command calls the PowerCLI snap-in to ensure that even if the script is run outside of the PowerCLI interface, it can still reference all the commands in the snap-in. As of Version 6 of PowerCLI, the `get-module -ListAvailable` command will now show the PowerCLI modules.

Writing the script

Writing a script should be straightforward and to the point. There needs to be a purpose, a breakdown of what inputs are needed, what outputs are needed, and who needs to run it. In most cases, this can be done on the fly as each line is typed, but it still is a methodized approach. This section deals with the mentality of writing a command, how to approach it, and planning the writing of it. As the IT Administration field looks more and more to DevOps frameworks, the more important building a process on how to do it becomes.

Planning the purpose

The basics of planning are to set the final expectations before beginning any typing.

Let's start with the example shown in the introduction of *Chapter 1, An Introduction to Essential Administration with PowerCLI*, where the boss needs a report of configuration data for a C-level individual in your company. What would a C-level be interested in seeing? Would it be CPU utilization and storage performance? Probably not—but will probably want a basic layout of what is in place. Let's start with a count of VMGuests in your environment.

Thinking through a script

So the purpose is to determine the number of VMs in the environment, so that will be the output. Wherever possible, use programmable methods to gather information as it ensures consistency and provides output that has limited user interaction to rope in information about the environment. In other words, try and make the script do the gathering of information instead of the end user.

Generally, for this type of script, we wouldn't necessarily need this much documentation but in these examples we will ensure this for consistency of thought. Let's start with the purpose:

```
<#
.Purpose or .Synopsis
Get the amount of VMs in the environment
```

Next is the description or detail of the script:

```
.Description or .Details
The environment consists of three vCenter's, The script needs to connect
to the three Management servers and poll all the VMs within all the
clusters. The output needs to get the total number of VMs and send the
output through to Boss@company.com
```

Is any input needed for the script? As all data will be captured as static entries, there is no input:

```
.Input
  None
```

What is the output for the script? The output is the final report with the VM counts:

```
.Output
  Report of total VM count
```

Lastly, for the information of the writer, use the following:

```
.Author
  Chris Halverson
.Change Log
  9/6/2015 - Initial Draft - Chris Halverson
.Filename
  Report-VMCount.ps1
```

```
.Version
0.1 Draft
#>
```

This is the script header; it helps put the scriptwriter into the mindset of what is going to be done in the script. It is very important to document this as it simplifies the process of figuring out what this script does by the writer several months later or through the eyes of the next DevOps administrator in the company. Once this is complete for the initial documentation, the next step should be to consider the inputs and what systems the script should connect to. As the script begins, it should start with some key commands such as connecting to the area that needs to be collected. For example, getting a Windows Management Interface connection starts with this:

```
Get-WMIobject
```

Connecting to vCenter begins with this:

```
Connect-VIServer
```

Connecting to the registry starts with this:

```
Get-ChildItem HKLM
```

All the environments use some sort of mechanism to connect to the desired subsystem. PowerShell modules are written to interface with something, and any script that is written has this premise in mind. The need is to connect to the VMware Automation Core component and then, the three vCenters are keys for the predetermined subsystem to connect to:

```
Add-PSSnapinVMware.VimAutomation.Core -ErrorActionSilentlyContinue
Connect-VIServer vCenter1, vCenter2, vCenter3
```

Once connected, plan how the information is to be collected. PoSH works in objects, as all **object oriented programming** (**OOP**) does. As Wikipedia defines, an object refers to a particular instance of a class where the object can be a combination of variables, functions, and data structures. So, if there is an examination of `Get-VM`, for instance, it produces an object. Porting that object to `$variable` shows a multi-level instance view of the VM. `$variable = Get-VM` pulls all the VMs in the environment into that `$variable`. This is a complex thought to process but it is critical to understand the information. Due to this information being captured in `$variable`, there can be an expansion of information and transferring into a data field into the output that is seen later in this chapter.

Planning the output

The last component of a script is determining how the output will be displayed and whether there is a delivery method for the script. Output to the screen is generally understood as a script being developed and while it can be useful, it isn't always desirable. Some outputs include **Comma Separated Value** (**CSV**), standard text file (TXT), **Out-GridView**, e-mail, HTML, and many others.

Typically, the CSV format plans for parsing and reintegration into other scripts. This is normally displayed through a spreadsheet and can be manipulated for detailed reports.

A TXT file is normally used for storage in a file sharing platform, displaying point-in-time configuration data. It normally is harder to parse and isn't used as commonly for reintegration for other scripts.

Out-GridView is a PowerShell-only output method that allows a nicer method to view table data.

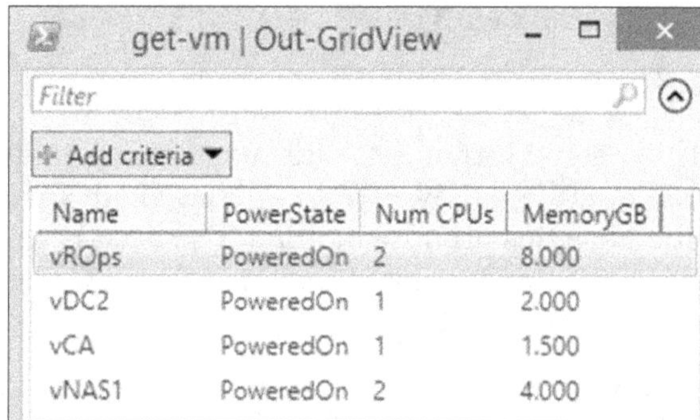

It displays an array of information in a readable format but is generally used as an administrator type output.

E-mail is normally the ideal output for automated reports. If the raw output data is sufficient for the receiver of the report, then this is the best type of output. This is also ideal for this report as it is simple and to the point.

Making the script

It turns out that this can be a one-liner and is written fairly easily. To understand the process, it is started using a key cmdlet and built upon that command. To get the list of VM guests, you run the `Get-VM` command, and it produces a list with these headings or output:

- `Name`
- `ConnectionState`
- `PowerState`
- `NumCPU`
- `CpuUsageMhz`
- `CpuTotalMhz`
- `MemoryUsageGB`
- `MemoryTotalGB`
- `Version`

This is all very useful information but too much for an example. So, we trim it with a select statement, such as `Get-VM | Select Name` or `Get-VM | % Name`, and it produces a simple but long list of VM names in the environment; however, the need is for a simple count generally using the `Count` method to produce a simple output:

```
$(Get-VM).Count
```

The output is as follows:

```
172
```

It is easy to add more to the output with `Write` statements and other beautification areas, but the simple number is usually enough. However, powered on VMs are more useful than powered off ones, so we add the following to the one-liner:

```
Write "The Powered On VM Count is: " $(Get-VM | where PowerState -match
"on").Count
```

The output is as follows:

```
The Powered On VM Count is:
160
```

Lastly, automate the sending of the information in an e-mail (note this is a single line):

```
Send-MailMessage -SMTP "SMTP.Company.com" -To "Boss@company.com" -From
"Admin@company.com" -subject "Current count of Powered On VMs" -Body
$(Get-VM | where PowerState -Match "On").Count
```

The output is as follows:

```
Sends an email with subject and information shown
```

Final script

The final script now looks like the following:

```
<#
.SYNOPSIS
Get the amount of VMs in the environment


.DESCRIPTION
The environment consists of three vCenter's, The script needs to connect
to the three Management servers and poll all the VMs within all the
clusters. The output needs to get the total number of VMs and send the
output through to Boss@company.com


.INPUT
  None
.OUTPUT
  Report of total VM count


.AUTHOR
  Chris Halverson


.CHANGE LOG
  9/6/2015 - Initial Draft - Chris Halverson


.FILENAME
  Report-VMCount.ps1


.VERSION
0.1 Draft


#>


# CONNECT TO VMWARE SNAPIN
```

```
Add-PSSnapinVMware.VimAutomation.Core -ErrorActionSilentlyContinue

# CONNECT TO vCENTERS AND DISPLAY NO CONNECTION DATA
Connect-VIServer vCenter1, vCenter2, vCenter3 | out-null

# START OF SCRIPT
Send-MailMessage -SMTP "SMTP.Company.com" -To "Boss@company.com" -From
"Admin@company.com" -subject "Current count of Powered On VMs" -Body
$(Get-VM | where PowerState -Match "On").Count

# DISCONNECT FROM vCENTERS WITHOUT CONFIRMATION
Disconnect-VIServer * -confirm:$false

# END SCRIPT
```

It seems long-winded for such a simple script and most would ask, "Why make this script so long when essentially the script block is one line?" Placing the script in this format allows for reuse, organization, further enhancements, and can be placed in a scheduled task or job at a later date. Updating and improving the script is key to growing in the art and making usable scripts reusable.

Report-VMHostConfigStatus

VMware created a method to maintain and keep a VMHost at a certain compliance level using a component called **Host Profiles**. Host Profiles maintain a record within vCenter to a base configuration using a **Reference Host**, and PowerCLI can be used to configure this reference host.

Running a simple PowerCLI command allows the creation of a basic profile:

```
New-VMHostProfile -Name TestProfile -Description "PowerCLI basic
Profile" -ReferenceHost VMHost1
```

It takes a few moments and a TestProfile is created within the vCenter environment. Default settings store domain authentication and the network and storage settings. It leaves the rest of the profile open for changes and alterations. As a consultant at numerous organizations, many companies feel the host profiles are overkill to a basic implementation script of certain configurations. So, with that in mind, let's think of implementing this type of script and name it `Report-VMHostConfigStatus.ps1`.

Starting with the inputs

Thinking this through, let's configure the script to examine the same attributes as the simple VMHost profile, with configuring the networking, storage, and basic host configurations such as **Network Time Protocol** (**NTP**), and DNS. So, to start with, we need the following details for the script:

- **Inputs**:
    - Hostname
    - Filename to store configuration information
- **Assumptions**:
    - NTP configuration will be an external time source of `0.ca.pool.ntp.org`, `1.ca.pool.ntp.org`, and `2.ca.pool.ntp.org`
    - The DNS will be 10.1.1.7, 10.1.1.8

Getting the current configuration

The following details are acquired using manual information gathering for Host NTP settings

1. DNS (estimated time 30 seconds per host):

   Select **Host** | **Configuration** | **DNS and Routing**.

   For vSphere 6.0 VIClient, you'll see the following output:

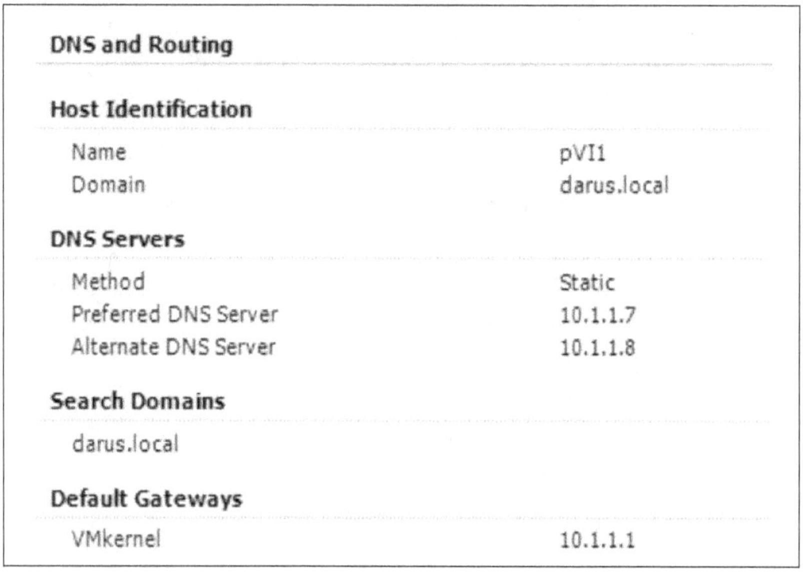

DNS and Routing

Host Identification

Name	pVI1
Domain	darus.local

DNS Servers

Method	Static
Preferred DNS Server	10.1.1.7
Alternate DNS Server	10.1.1.8

Search Domains

darus.local

Default Gateways

VMkernel	10.1.1.1

Select **Host** | **Manage** | **TCP/IP configuration** | **DNS**.
For vSphere 6.0 Web Client, you'll see the following output:

DNS | Routing | IPv4 Routing Table | Advanced

Configuration method:	Use manual settings
Host name:	pVI1
Domain:	darus.local
Preferred DNS server:	10.1.1.7
Alternate DNS server:	10.1.1.8
Search domains:	darus.local

2. NTP (estimated 15 seconds per host):

   Navigate to **Host** | **Configuration** | **Time Configuration**.

   For vSphere 6.0 VIClient, you'll see the following output:

Time Configuration

General

Date & Time	19:46 9/10/2015
NTP Client	Running
NTP Servers	0.ca.pool.ntp.org, 1.ca.pool.ntp.org, 2.ca.pool.ntp.org

Select **Host** | **Manage** | **Time Configuration**.

For vSphere 6.0 Web Client, you'll see the following output:

Time Configuration

Date & Time:	9/10/2015 7:44 PM
NTP Client:	Enabled
NTP Service Status:	Running
NTP Servers:	0.ca.pool.ntp.org, 1.ca.pool.ntp.org, 2.ca.pool.ntp.org

3. Record network configuration Standard Switches (with CDP configuration) (estimated 10 minutes per host):
   Select **Host** | **Manage** | **Virtual Switches**, and click on the (i) icon:

   For vSphere 6.0 Web Client, you'll see the following output:

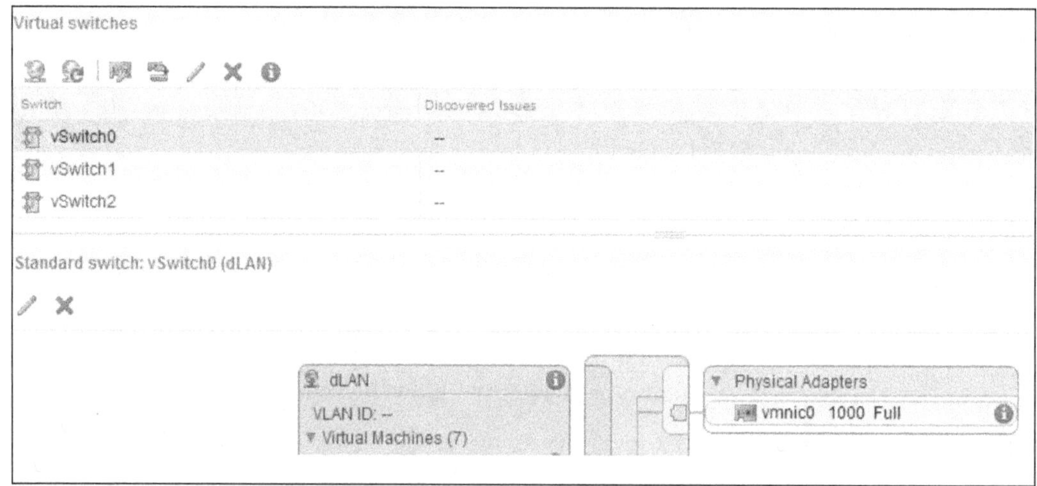

Now, select **CDP**:

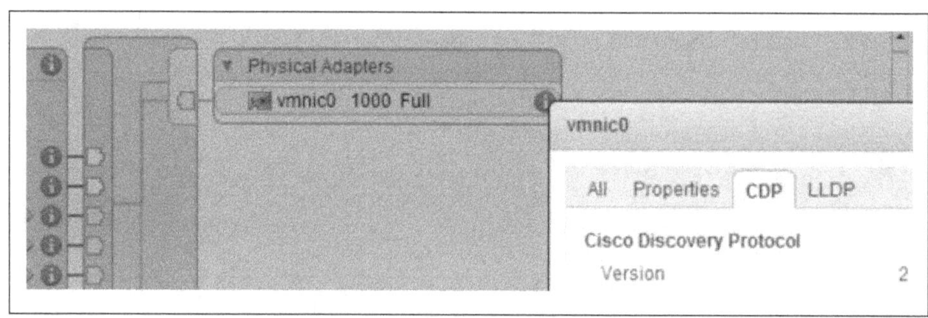

4. Record Storage LUNs and NFS share configuration (estimated 15 minutes per host).

   For vSphere 6.0 VIClient, you'll see the following output:

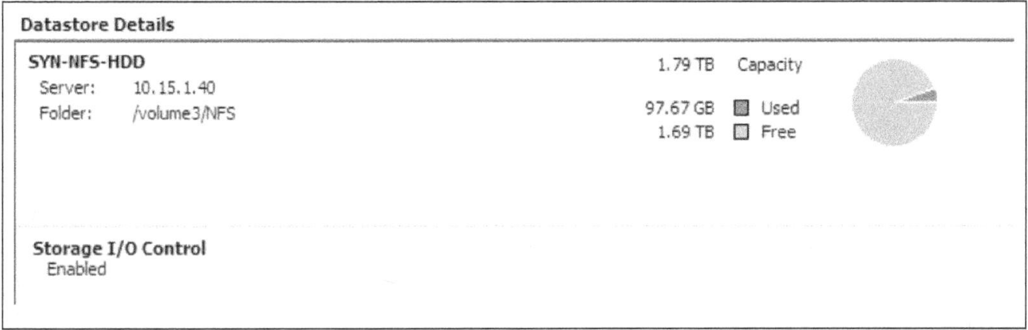

These screenshots are a sample of some of the information to be gathered and they are certainly not exhaustive. Each component suffers from possible inconsistent recording of data as the recorder may inadvertently jot down an incorrect or case-insensitive value devaluing the collected data.

Output of data

Once the data is collected, an output is generally chosen that will allow all the information to be shown easily. Collection of data up to now has been manual, and thus, this report will also have to be manually generated. Typically, the output is put into a spreadsheet or with a Rich Text Format type of word processor. This is the typical time-consuming part, as one that has had to do this numerous times.

Getting the same information through PowerCLI

To set up the documentation, use the following code:

```
<#
.SYNOPSIS
  Get the VMhost information from the entire environment


.DESCRIPTION
```

```
  The environment consists of three vCenters and this script needs to
connect to the three Management servers and poll all the VMhosts within
all the clusters. The output needs to be to a spreadsheet or a CSV.


.INPUT
  None
.OUTPUT
  CSV or Excel spreadsheet


.AUTHOR
  Chris Halverson


.CHANGE LOG
  9/16/2015 - Initial Draft - Chris Halverson


.FILENAME
  Report-VMHost.ps1


.VERSION
  0.1 Draft

#>
```

Next is the connection to the VMware/PowerCLI subsystem:

```
# CONNECT TO VMWARE SNAPIN
Add-PSSnapinVMware.VimAutomation.Core -ErrorActionSilentlyContinue


# CONNECT TO vCENTERS AND DISPLAY NO CONNECTION DATA
Connect-VIServer vCenter1, vCenter2, vCenter3 | out-null
```

Now that the vCenters are connected to, the script should collect the names of the VMHosts:

```
$vHost = Get-VMhost
```

Do we need to capture the vCenter first and then the VMHost that resides within the vCenter or does it matter if they are all lumped together? In this case, we will lump them together. Naming the VMHosts shows a site identifier in the name (for example, `YYC-host01`), and determining the naming afterward will be simple or irrelevant.

`$vHost` now contains all the host information and certain information can now be gathered from it. So the output of the `$vHost.name` command is as follows:

```
SJC-Host01
SJC-Host02
SJC-Host03
SJC-Host04
SJC-Host05
SJC-Host06
SJC-Host07
SJC-Host08
SEA-Host01
SEA-Host02
SEA-Host03
SEA-Host04
SEA-Host05
SEA-Host06
SEA-Host07
SEA-Host08
YYC-Host01
YYC-Host02
YYC-Host03
YYC-Host04
```

Otherwise, the output for the following command would be total pCPU cores:

```
ForEach ($nCPU in $vHost.NumCPU) {$NumCPU = $NumCPU + $nCPU}
```

So, assuming that each host had two sockets with 8 cores apiece, the output would be (*16 cores per host * 20 hosts = 320 pCPUs*) 320 pCPUs. This can assist in a capacity report but won't be used here.

Domain Name Service (**DNS**) is gathered from various interfaces on the host; therefore, the information has to be captured in a per interface fashion. It turns out `$vhost | Get-VMhostNetwork` provides the information needed. There is a DNSAddress object within the `Get-VMHostNetwork` command. So pulling out a command such as `$($vhost | Get-VMHostNetwork).DNSAddress` provides the primary and secondary DNS servers as a two line entry, 10.1.1.7 and 10.1.1.8 for each VMHost.

Network Time Protocol (**NTP**) is essential to maintaining proper time synchronization within an environment and is crucial to the authentication mechanism within Active Directory for any security correlation and centralized logging. NTP info is simple to capture with this specific command, `$vHost | Get-VMHostNTPserver`. A blank configuration generally means it is not configured, and therefore, depends on an inaccurate computer clock for log synchronization.

We could use a `ForEach` loop to cycle through the hosts and pull the DNS and NTP information:

```
ForEach ($Item in $vHost) {
  $Item.Name
  $($Item | Get-VMhostNetwork).DNSAddress
  $Item | Get-VMhostNTPServer
}
```

This provides the information needed for the required script, but it isn't pretty, but hey, it is the output.

The following is the output:

```
SJC-Host01
10.1.1.7
10.1.1.8
0.pool.ntp.org
1.pool.ntp.org
2.pool.ntp.org
SJC-Host02
10.1.1.7
10.1.1.8
0.pool.ntp.org
1.pool.ntp.org
2.pool.ntp.org
SJC-Host03
```

```
10.1.1.7
10.1.1.8
0.pool.ntp.org
1.pool.ntp.org
2.pool.ntp.org
SJC-Host04
10.1.1.7
10.1.1.8
0.pool.ntp.org
1.pool.ntp.org
2.pool.ntp.org
SJC-Host05
10.1.1.7
10.1.1.8
0.pool.ntp.org
1.pool.ntp.org
2.pool.ntp.org
SJC-Host06
10.1.1.7
10.1.1.8
0.pool.ntp.org
1.pool.ntp.org
2.pool.ntp.org
SJC-Host07
10.1.1.7
10.1.1.8 ...
```

As seen, the output is long-winded and repetitive; however, it is informative; two hosts in Seattle are without an NTP server setting and are likely not turned on, and a couple machines have `time.nist.gov` configured instead of `0.pool.ntp.org`. So it is known what the appropriate values should be and changing the code to show differences instead of listing everything through an additional pipeline command can be added to show only the anomalies:

```
ForEach ($Item in $vHost) {
  $Item.Name
  $($Item | Get-VMhostNetwork).DNSAddress
```

```
    $Item | Get-VMhostNTPServer | where { $_ -notmatch
      "pool.ntp.org"}
}
```

The `$_` symbol refers to the direct output before the pipeline and the output only shows the match to query `notmatch "pool.ntp.org"`, it will still show the VMhosts that have no NTP and the misconfigured in the list. The script can produce a method to perform corrective action upon a misconfiguration, but we desire a report and not a rectification engine, so the initial script can be used.

Network

Every host has a virtual switch that contains the interface that connects the host to the vCenter management and provides a portgroup for VMGuest Network traffic. Whether Enterprise Plus is licensed and Distributed vSwitches is used or Standard licensing is in place and standard vSwitches are assigned, all hosts have virtual networking. Capturing the information is a simple enough process with `$vHost | Get-VirtualSwitch` and it lists both Standard and Distributed.

What information is important here? The information that can be gathered is huge but the original intent was that this could essentially replace a host profile, so the following information needs to be recorded:

- **vSwitch**

  Name

  `$($vHost | Get-VirtualSwitch).Name`

  Number of Ports

  `$($vHost | Get-VirtualSwitch).Numports`

- **PortGroup**

  Name

  `$($vHost | Get-VirtualSwitch | Get-VirtualPortGroup).Name`

  VLAN ID

  `$($vHost | Get-VirtualSwitch | Get-VirtualPortGroup).VLANID`

  Interface attached

  `$($vHost | Get-VirtualSwitch).Nic`

  MTU

  `$($vHost | Get-VirtualSwitch).MTU`

Distributed or Standard

```
$vHost | Get-VirtualSwitch -Distributed
```

```
$vHost | Get-VirtualSwitch -Standard
```

- **Virtual Management interface**

```
$vHost | Get-VMHostNetworkAdapter -Vmkernel | select Name,
PortGroupName, IP, SubnetMask, MTU, ManagementTrafficEnabled,
vMotionEnabled, FaultToleranceLoggingEnabled, VSANTrafficEnabled
```

Take notice that these one-liners are certainly not the only way to get this information. These are but a listing of an example method, or a single process to get the information.

Storage

Storage reporting primarily depends on a few factors but doesn't usually display in one specific place. Normal reporting for a file-based storage (NFS) would include IPs for the Storage, path, and the VMHost network to name a few of the components, whereas, block-based storage (Fiber Channel, iSCSI) has different criteria such as LUN and HBA, and if Software iSCSI initiators are used, the iSCSI binding configuration.

So, following the same format from the *Network* section previously, it will start with the `Datastore` setup.

- **Datastore**

Name

```
$($vhost | get-datastore).Name
```

Type (NFS / VMFS)

```
$($vhost | get-datastore).Type
```

Capacity in GB

```
$($vhost | get-datastore).CapacityGB
```

Free Space in GB

```
$($vhost | get-datastore).FreeSpaceGB
```

Connection information

```
$DS = $vHost[0] | Get-DataStore | ? Type -eq "NFS"
```

```
$DS.Name
```

```
$DS.RemoteHost
```

Disk-specific information (looking at the first host)

```
$ds = $vHost[0] | Get-Datastore | ? Type -eq "VMFS"

$view = $vhost[0] | get-view

$view.config.storagedevice.scsiLun | ? DeviceName -match
$ds[0].extensiondata.info.vmfs.extent.diskname
```

Output

Originally, a CSV file was to be used but formatting for all the information would be difficult, as the amount of components would change. From the number of NICs to the amount of NTP servers in use, the CSV wouldn't work as well as a standard text report. With that being said, if a CSV were to be used for output, the initial line would start with the following code:

```
$Headers = "VMhost Name, NTP Server, DNS Server, Network Setting, Storage
Setting"


Write "$Headers" | Out-File -Encoding Ascii $CSVName -Force


$CSVOut = "$Item.Name, $Item.NTP, $Item.DNS, $Item.NetSet, $Item.Storage"


Write "$CSVOut" | Out-File -Encoding Ascii $CSVName-Force -append
```

Otherwise, the output will be just a standard `Out-File -FilePath <path>/filename.txt -Append` concluding any output statements in the code. The combination of all of these script snippets culminate into another example.

Final script of the report

The final script now looks as follows:

```
<#

.SYNOPSIS

Get the VMhost information from the entire environment


.DESCRIPTION

The environment consists of three vCenters and this script needs to
connect to the three Management servers and poll all the VMhosts within
all the clusters. The output needs to be to a spreadsheet or a CSV.


.INPUT
```

```
  None
.OUTPUT
  CSV or Excel spreadsheet

.AUTHOR
  Chris Halverson

.CHANGE LOG
  9/16/2015 - Initial Draft - Chris Halverson

.FILENAME
  Report-VMHost.ps1

.VERSION
0.1 Draft

#>

# CONNECT TO VMWARE SNAPIN
Add-PS SnapinVMware.VimAutomation.Core -ErrorAction SilentlyContinue

# CONNECT TO vCENTERS AND DISPLAY NO CONNECTION DATA
Connect-VIServer vCenter1, vCenter2, vCenter3 | out-null

# Define the path and filename of the final output
$fileOutput = "c:\scripts\output\Host-Report.txt"

# Get the VMhosts into an array
$vHosts = Get-VMHost

# Cycle through the Hosts and get the required information

foreach ($vhost in $vhosts) {
  "Host Information" | Out-File -FilePath $FileOutput

#Display Name
```

```
Write "Name: " $vhost.Name | Out-File -filepath $fileOutput -Append


#Display DNS
Write "DNS: " $($vHost | Get-VMhostNetwork).DNSAddress | Out-File -
filepath $fileOutput -Append


#Display NTP servers
Write "NTP: " $vHost | Get-VMhostNTPServer | Out-File -filepath
$fileOutput -Append


Write"Networking Information" | Out-File -FilePath $fileOutput -Append
$vHost | Get-VMHostNetworkAdapter -Vmkernel | select Name,
PortGroupName, IP, SubnetMask, MTU, ManagementTrafficEnabled,
vMotionEnabled, FaultToleranceLoggingEnabled, VSANTrafficEnabled |
Out-File -filepath $fileOutput -Append


#Display vSwitches
Foreach ($vSwitch in $($vHost | Get-VirtualSwitch)) {
Write "Name: " $vSwtich.Name | Out-File -filepath $fileOutput -Append
#Display Number of Ports
Write "Number of Ports: " $vSwitch.NumPorts | Out-File -filepath
$fileOutput -Append
#Display Attached pNICs
Write "pNICs: " $vSwitch.NIC | Out-File -filepath $fileOutput -Append
#Display MTU
Write "MTU: " $vSwitch.MTU | Out-File -filepath $fileOutput -Append


#Attached PortGroups
Foreach ($pg in $vSwitch) {
  Write "Name: " $pg.Name | Out-File -filepath $fileOutput -Append
  Write "VLAN: " $pg.vlanid | Out-File -filepath $fileOutput -Append
  }
}
  #Display Storage Information
  "Storage Information" | Out-File -FilePath $FileOutput -Append


  Foreach ($DS in $($vHost | Get-Datastore)) {
```

```
    Write "Datastore Name: " $DS.Name | Out-File -filepath
    $fileOutput -Append
    Write "Type: " $DS.Type | Out-File -filepath $fileOutput -Append
    If ($DS.Type -eq "NFS") {
      Write "NFS Connection Info: " $DS.RemoteHost | Out-File
      -filepath $fileOutput -Append
    }
    Write "Capacity GB: " $DS.CapacityGB | Out-File -filepath
      $fileOutput -Append
    Write "FreeSpace GB: " $DS.FreeSpaceGB | Out-File -filepath
      $fileOutput -Append
    $view = $vHost | Get-View
    $view.config.storagedevice.scsiLun | ? DeviceName -match
    $ds.extensiondata.info.vmfs.extent.diskname| Out-File -filepath
    $fileOutput -Append
  }
}
```

Although the output isn't as pretty, the data is displayed in a consistent and simple flow. The `Output` file is ready for printing, e-mailing, or filing in a file share.

Summary

In this chapter, you learned about tools to help in the development cycle, how to work through a point and click administration conversion to a PowerShell script, and developed a simple script to perform such tasks. In *Chapter 3, Enhancing the Scripting Experience*, we will explore how to enhance the scripting experience, develop a repository, and adapt functions and calls of other scripts to improve the simplicity of the scripts.

3
Enhancing the Scripting Experience

Monday morning rolls around and the boss comes into your office with a big smile on his face, says "Great report!" in his loud booming voice, and continues, "The information you provided helped the department secure another $150k for the budget. We are supposed to build a new automation framework that will allow us to produce these types of reports for the management team on a weekly basis." So you reply, "So we got a $150k budget for developing this framework in-house?". The boss chuckles and says, "No, it is for an office renovation. I'm getting my corner window office!" and he walks away laughing. "Typical!" you think.

In building an automation framework, there needs to be a methodology to document procedures, create central stores, establish tools, design development standards, and form repositories. Each piece takes time, and in some cases, capital. However, setting precedence beforehand makes the task of sorting and running the scripts later handy and usable. This chapter deals with building a central script store and establishing a means to build consistent and usable scripting experience for provisioning, reporting, or configuring a virtual environment simple and straightforward. We will also deal with:

- Building a code repository
- Scripting with the intention of repurposing
- Establishing quality control

Code repositories – why would I need that?

The first thing when developing code, by spending hours designing and writing the script, is to find the script afterwards. I, personally, have spent days writing a script, to do something that I feel will be useful, and placed it in a folder on my computer and then have completely forgotten about where it is when I need it again 3–6 months later. Another example of this is when I was in the midst of writing a script just to find out a work associate is writing the same thing, and lastly, I pass along a script that I have worked on for a few days to another member on the team, they attempt to run the script, and respond, "It doesn't work."

Having a well-defined repository and a good definition of the script gives all who may be developing in it an open and robust place that everyone on the team can build upon. A repository can consist of a shared folder on a centralized file share, a document share such as SharePoint, a commercial code repository such as CVS or Team Foundation Server, or a public-facing code sharing site such as personal **Software Change Manager** (**SCM**), or even a personal blogsite. The value of the repository is dependent upon the sensitivity of the script data being placed in the store. If the scripts are being written with the intention of sharing with a work associate, a file share may be enough; however, if the scripts are very generic and shares no company-sensitive data, and there is a desire to get more feedback and more contributors, a public SCM account might be beneficial.

Building an internal repository

Let's start with a file share layout. Most IT shops have a central server that tends to be a "Jump box" or utility server that runs general scripts or contains a multitude of administration tools. This server could be set up with the file folder named `Scripts`, which will contain the repository.

```
PowerCLI C:\> mkdir Scripts

    Directory: C:\

Mode                LastWriteTime     Length Name
----                -------------     ------ ----
d----        10/28/2015   8:22 PM            Scripts

PowerCLI C:\> _
```

Next, let's think through the structure. Any good store of scripts should have the following parameters:

- An area that stores final scripts
- An area that stores scripts under development
- A way to label versions
- A method for having a runtime area that has input files and output files.

All of this enables consistency, provides a means to develop more functionality in the code, helps control the script's content, and allows additional functionality to be developed while still allowing the script to be run.

So with that in mind, this is an example of a folder structure for `Scripts`:

```
Scripts
    Final
        Input
        Output
        Modules
    Test
        Input
        Output
        Modules
    Development
        Input
        Output
        Modules
```

Storing the scripts in the main folder will allow a repository to be useful in layout. Start by building the scripts in `Development`; once the script is functional, a copy will be placed in the `Test` folder and will be available for testing by the remainder of the operational staff. Once confirmed, the script would be copied to the `Final` or `Production` folder. This would allow multiple copies of a script set for different stages of the code base.

The main disadvantage of this is that it quickly becomes unruly and loses its effectiveness once the script repository exceeds a couple hundred scripts, or the user/developer base exceeds about 10 individuals. This is mostly due to the check-in/check-out process. It is very manual and prone to overwrites and loss due to the process. Other issues include having multiple people trying to edit the same script at the same time, problems when the versions become out of sync (editing version 1.4, while the script that is in test is version 1.1 and the one in Production is version 1.3), and not knowing what the latest version to work with is. The fact is that there is no way to lock a script and see its history, and this can be one of the biggest issues.

Using a third-party tool for code check-in and check-out

Downloading and integration aside, there are numerous solutions that could be used to help alleviate the check-in/check-out problem. Using a document management solution such as Microsoft SharePoint or EMC Documentum can resolve some of the version control and sign in and sign out issues; however, these can be large and expensive solutions requiring local support staff to use and this truly is a relatively small issue. If these tools are already running in the environment, look at them as an option for the `Development` portion of the repository.

The alternative is utilizing a **Version Control System** (**VCS**) or **Software Change Manager** (**SCM**), which is smaller, has a number of open source forks, and can solve a number of the same issues. As with any solution, there is the typical problem of another product to learn and another product to support.

Any solution that is to be installed in the site needs to be robust and jointly agreed on by all parties that will use it; `Development`, `Administration`, and `Users` of it. Having one individual that supports and operates a proprietary solution will become either a bottleneck or will result in losing the point of the original solution when that person finds a different job.

Send it to the cloud! Publicly accessible script repositories

There are a number of places to store code snippets, one-liners, and short examples for the team to view and comment on. Blog sites such as Blogger or WordPress can provide a static means of storing scripts and wiki can provide a simple collaboration platform for the team. Then, share these items as a team within this location and the information can be useful. It is a static area and for most individuals provides a place to share their experience.

GitHub is an example of a cloud-based VCS and provides up to a five-user repository for free. This site adds to the VCS that we talked about in the previous section and allows scripts to be collaborated on by a team that may not be resident in the same geographic area or in the same time zone.

Conclusions

No matter what method is chosen, the organization of scripts should be planned beforehand. Planning the structure helps ensure that whatever automation mechanism could be chosen in the future, such as VMware vRealize Automation, VMware vRealize Orchestration, or some other custom developed workflow, the scripts will be available to anyone requiring them to run.

Scripting with the intention to reuse

Changing gears from storing the scripts is the ability to repurpose them, or use portions of a script in multiple areas. Functions, workflows, and modules are critical to this end. Each will be described in detail in this section of the book.

Functions, functions, functions – the fun with functions

What is a function? The `help` file defines it as a named block of code that performs an action. So writing a function would look like this:

```
Function Get-PS {get-process PowerShell}
```

Here, the output would be nothing until the function is called like this:

```
Function Get-PS {get-process PowerShell}
Get-PS
```

The output of this is the same as `Get-Process PowerShell` but it is reusable and could be called at any point in the script. There is a simple and useful function that I personally use in many of the developed scripts to get a decision and return `true` or `false`:

```
Function Get-Decision ($Decision) {
  Switch ($Decision.toupper()) {
    "Y" {$Decision = $True}
    "N" {$Decision = $False}
```

```
    Default {"Enter a ''Y'' for a Yes or a ''N'' for No." ;
        Get-Decision ($Decision)}
    }
  Return ($Decision)
}
```

As you can see, the function is called `Get-Decision`, accepts an object through `Read-Host`, and returns the decision. So the command that would call this would be something like this:

```
$MakeCSV = Get-Decision $(Read-Host "`nDo you need to make this a
CSV to duplicate the settings and create more than one? (Y/N)")
```

So the `$MakeCSV` variable would be either `$True` or `$False`, depending on the outcome of the user prompt. Then, the next line would do something based upon that decision.

All functions *must* be before the mainline code. Every PoSH runs in a linear fashion from start to end; therefore, if the function is being called, it has to have been run to enter it into running memory. The function is also removed from the memory upon closing the command window, so think through where the function is needed and open it accordingly.

Example of a multifunctional script

Consider the following parameters:

- **Premise:** Build a script that automates the building of a VM
- **Inputs:** The script should capture the VM name, list how many CPUs and how much memory is required, and then build a VM with the settings captured and a 50 GB hard disk
- **Output:** This builds the VM
- **Assumption:** `$cred` is captured using the `$cred = Get-Credential` command before this script is run
- **Script**: Consider the following code snippet:

```
If ($Cred -ne $null)
    {
    Connect-ViServer vCenter -credential $cred | out-null
    }
Else
    {
```

```
    End
    }


Function Gather-Data
{
$NewVM= @{
    Name = Read-Host "What is the name of the VM?";
    NumCPU = Read-Host "What is the amount of CPUs
        required?";
    MeminGB = Read-Host "How much RAM is required?"
    VMHost = get-vmhost | select -first 1
    }
  Return ($NewVM)
}


Function Build-VM ($NewVM)
  {
    New-VM -Name $NewVM.Name -NumCPU $NewVM.NumCPU `
        -MemoryGB $NewVM.MeminGB -DiskGB 50 -vmhost `
            $NewVM.vmhost
    }


$output = Gather-Data
Build-VM($output)
```

This is a pretty simple example of building a hash table (also referred to as an array) where it maps keys to a value in one function, outputting that hash table and using it as the input for a second function. The information stored in the hash table is used to populate a `New-VM` command. The point is that the `Gather-Data` function can be used to collect data for other things such as a CSV file, or a different function that searches for the information and removes the VM. The best part is that the information is deleted once the window is closed and is useful only when the script is run.

Using modules

A module is, again as pointed out in the `help` file, a package of commands. It is a collection of cmdlets, scripts, workflows, aliases, variables, and functions that when imported, allow the user to run `true` and tested commands. Every installation of PoSH has modules installed by default. To view the installed modules, type `Get-Module -ListAvailable` and the command will list the installed modules. After the installation of the PowerCLI package, there are a number of modules that are added to the available set of modules:

```
Directory: C:\Program Files (x86)\VMware\Infrastructure\vSphere PowerCLI\
Modules


ModuleType Version    Name                                ExportedCommands

---------- -------    ----                                ----------------

Binary     6.0.0.0    VMware.VimAutomation.Cis.Core

Binary     6.0.0.0    VMware.VimAutomation.Cloud

Manifest   6.0.0.0    VMware.VimAutomation.Core

Binary     6.0.0.0    VMware.VimAutomation.HA

Binary     1.0.0.0    VMware.VimAutomation.License

Binary     6.0.0.0    VMware.VimAutomation.PCloud

Manifest   6.0.0.0    VMware.VimAutomation.SDK

Binary     6.0.0.0    VMware.VimAutomation.Storage

Binary     6.0.0.0    VMware.VimAutomation.Vds

Binary     1.0.0.0    VMware.VimAutomation.vROps

Binary     6.0.0.0    VMware.VumAutomation
```

Each of these modules includes numerous pieces that perform various functions with the VMware environment. To list the commands available in a module, for example `VMware.VimAutomation.Core`, run the command `Get-Command -module VMware.VimAutomation.Core` where it outputs 291 individual cmdlets as of PowerCLI v6.0 R3.

With the use of modules, and using PoSH v3, importing modules from another system is now a reality. This importing isn't a permanent addition but can be very handy in a pinch. Firstly, WinRM must be running and configured for remote management; this is done through the command `WinRM quickconfig`, running on an administrative PowerShell window. Once WinRM allows remote connections, type `$PS = new-pssession -computername <computername>`. This allows a remote connection through WinRM to the other computer. Once this `PSSession` is established by running the following command, commands can be run on the alternate computer and we can import modules to the local computer and run them.

```
$PS = New-PSSession -ComputerName TestComputer
Get-Module -PSSession $PS -ListAvailable
```

This shows the available modules on the remote computer. Adding the following line of code will permit the import of the module into the current PoSH window:

```
Import-Module -PSSession $PS Module_name
```

So if there is no `VMware.VimAutomation.Core` installed on the local system, this method allows it to run as if it were installed. The running of the command is slower due to having to run it through the remote connection, but it does run and can get the information needed. Note that it doesn't transfer the aliases as well, so try and remember the full syntax for the command.

Building modules

As building modules are a little more of an advanced topic, this book will only discuss the use of them. However, a little understanding of what they are may whet the appetite of the reader to delve deeper into the topic. MSDN outlines in a good read about four different module types: the script module, binary module, manifest module, and dynamic module:

(`https://msdn.microsoft.com/en-us/library/dd878324%28v=vs.85%29.aspx`)

Primarily, PowerCLI modules have always been typically plugins and not modules; they always required the separate running of the PowerCLI icon to allow the use of their cmdlets. Recent versions have changed that mantra and are now using modules. Type `Get-Help About_PSSnapins` in your PowerShell interface to learn more about the difference.

Calling other scripts within the script

Calling other scripts within a script is fairly straightforward as it is a single command: `Invoke-Expression -command .\script.ps1` or `"c:\scripts\sample_script.ps1" | invoke-expression`. This allows for an external script to be run and for the output of that script to be captured within the main script. Think about it this way; as the wrapper, `ps1`, or a scripting framework that has been developed, it can call another script that another person can write, and the output from the external expression. It wouldn't matter what the output would need to be unless the output needs to be parsed and processed for a different output.

Use the example that was written in the multifunctional section previously—there is a need to capture the VM name, know how many CPUs and how much memory is required, and then build a VM with the settings captured and a 50 GB hard disk.

Example of a framework script

Consider the following parameters:

- **Premise**: Build a script that automates the building of a VM
- **Inputs**: The script should capture the VM name; list how many CPUs and how much memory is required, and then build a VM with the settings captured and a 50 GB hard disk
- **Output**: This builds the VM
- **Assumption**: This script must call other scripts through `invoke-expression`
- **Script**: Consider the following code snippet:

  `Connect-vCenter.ps1`

```
<#
.Synopsis
    Does the lifting of connecting to a vCenter
.Description
    Starts VIMAutomation Module
    Gets Credentials
    Gets the vCenter name and connects to vCenter
.Input
    User Input
        Credentials
        vCenter Name
.Output
```

```
    None
.Author
    Chris Halverson
.Change Log
    11/6/2015
.FileName
    Connect-vCenter.ps1
.Version
    Draft 0.1
#>

#Call the Module
If ($(get-module -ListAvailable) -notmatch `
"VMware.VimAutomation.Core") {
    Import-Module VMware.vimautomation.core
    }

#Test for stored credentials
If ($Cred -ne $null) {
    write-Host 'Using the Credential '$Cred.UserName
    }
Else {
    $Cred = Get-Credential
    }
$VC = Read-Host "What is the name of the vCenter server?"

#Trying a ping of the vCenter Host
If (Test-Connection -ComputerName $VC -Count 2 -quiet) {
    Connect-ViServer $VC -Credential $Cred | Out-Null
    Write-Host "Connected to vCenter $VC " -ForegroundColor
      Green
    [Console]::ForegroundColor = "Gray"
    }
Else {
    Write-Host "vCenter not found. Exiting script" `
    -ForegroundColor Red
```

```
    [console]::ForegroundColor = "Gray"
    Break}
```

Gather-Data.ps1

```
<#
.Synopsis
    Collects Information for build data
.Description
    Gets Appropriate data and outputs to a variable
.Input
    User Input
        VM Name
        Number of CPUs
        Amount of RAM in GB
.Output
    Variable
.Author
    Chris Halverson
.Change Log
    11/6/2015
.FileName
    Gather-Data.ps1
.Version
    Draft 0.1
#>

$global:Sys = @{
    Name = Read-Host "What is the name of the VM?";
    NumCPU = Read-Host "What is the amount of CPUs required?";
    MeminGB = Read-Host "How much RAM is required?"
    }
```

Build-VM.ps1

```
<#
.Synopsis
    Builds a VM
.Description
```

```
        Calls the Connect-vCenter.ps1 Script to connect to vCenter
        Gathers Data from Gather-Data.ps1
        Builds VM based on Specs
    .Input
        Called through other Scripts
    .Output
        Builds VM
    .Author
        Chris Halverson
    .Change Log
        11/6/2015
    .FileName
        Build-VM.ps1
    .Version
        Draft 0.1
    #>

    Invoke-Expression -command .\Connect-vCenter.ps1
    Invoke-Expression -command .\Gather-Data.ps1

    $vHost = Get-VMhost | Select -first 1

    New-VM -Name $Sys.Name -NumCPU $Sys.NumCPU -MemoryGB $Sys.MeminGB
    ` -DiskGB 50 -vmhost $vhost
```

Upon completion of the `Build-VM.ps1` script, a new VM is created and although the Guest OS type is Windows XP and some of the other configurations are a little off, like a typical build would be, the script was successful and it has components that can be reused in a loop or a workflow.

Building a framework that others can build upon

A framework in this context is a set of standard components that every other script developed would want to access, for example, building a wrapper that can do the e-mailing of a final output without running the command and specifying certain information, or something that automatically opens the connection to the vCenters in the environment. These are some of the things a wrapper or workflow can do.

Some of the advantages of workflows are that they can run in parallel instead of serially, as compared to a standard PowerShell script; they can be paused, stopped, or restarted as needed; they can be run on a remote hosts and can increase performance 10-fold if there are enough remote engines to run it and the logging is included in the workflow engine. These are huge, as more and more administration needs to have parallelism to run properly. One important thing to note is that workflows are built using the .NET Framework. The **Windows Workflow Foundation** (**WWF**) and the PowerShell code that are being run in the workflow are actually being translated into XAML for WWF, so it can run.

The XAML code can actually be seen when a workflow is created, by typing the following command line:

```
get-command [Workflow name] | select-object XamlDefinition
```

Typing `Help About_Workflows` gives a lengthy bit of information that discusses what a workflow is, its benefits, and why it is needed.

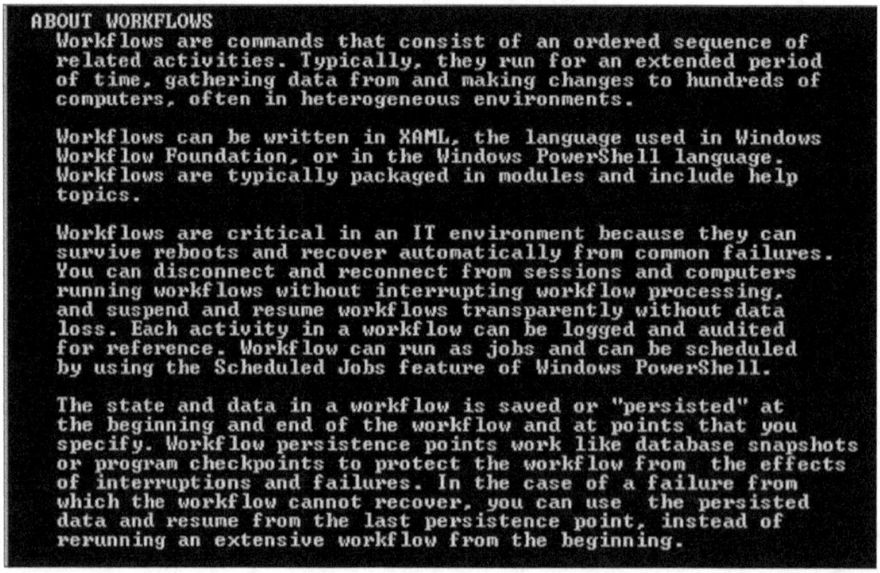

```
ABOUT WORKFLOWS
    Workflows are commands that consist of an ordered sequence of
    related activities. Typically, they run for an extended period
    of time, gathering data from and making changes to hundreds of
    computers, often in heterogeneous environments.

    Workflows can be written in XAML, the language used in Windows
    Workflow Foundation, or in the Windows PowerShell language.
    Workflows are typically packaged in modules and include help
    topics.

    Workflows are critical in an IT environment because they can
    survive reboots and recover automatically from common failures.
    You can disconnect and reconnect from sessions and computers
    running workflows without interrupting workflow processing,
    and suspend and resume workflows transparently without data
    loss. Each activity in a workflow can be logged and audited
    for reference. Workflow can run as jobs and can be scheduled
    by using the Scheduled Jobs feature of Windows PowerShell.

    The state and data in a workflow is saved or "persisted" at
    the beginning and end of the workflow and at points that you
    specify. Workflow persistence points work like database snapshots
    or program checkpoints to protect the workflow from  the effects
    of interruptions and failures. In the case of a failure from
    which the workflow cannot recover, you can use  the persisted
    data and resume from the last persistence point, instead of
    rerunning an extensive workflow from the beginning.
```

So to add to the description from the `help` file, these are jobs, or scheduled jobs that can be started, run, rerun, and paused for a reboot. The workflow engine should be using PoSH v3 or newer to run, as these jobs gain many pluses when run through the workflow engine. The workflow requires the following:

- A client computer to run the workflow
- A workflow session (a PowerShell session, otherwise known as a **PSSession**) on the client computer (it can be local or remote)
- A target for the workflow activities

Running a workflow

To run a workflow, there are some caveats to make sure things run smoothly. Firstly, make sure that PowerShell is running in an administrative window or option. Either right-click on the PowerShell icon and select **Run as Administrator** (which triggers an User Access Control or UAC verification window) or type `Start-Process PowerShell -verb RunAs` that does the same thing.

Next, enable the client or the remote client to have a remote PSSession run by typing `Enable-PSRemoting -Force` or `Set-WSManQuickConfig` or, as seen in a previous section, `WinRM QuickConfig`, which allows the computer to accept WinRM or remote management commands. These commands will start the WinRM service, configure a firewall exception for the service, and allow Kerberos authentication for the commands to be run.

To return the configuration back to its original condition

`Disable-PSRemoting -force` returns the configuration to the original condition of the OS.

The typical way to set up the remote session is to use the `New-PSWorkflowSession` cmdlet. Let's see what this actually looks like and then process a simple workflow script. Consider the following points:

- **Premise**: Build a workflow that gets a WMI entry for Win32_BIOS, lists running processes, and lists the stopped services on the computer, and does all that in parallel
- **Inputs**: All inputs will be hardcoded for simplicity
- **Output**: This includes the content of `Win32_BIOS`, processes that are running, and the stopped services
- **Assumption**: This indicates that a workflow is being used

- **Script**: Consider the following code snippet:

  Workflow

```
Workflow pTest {
  Parallel {
    Get-CimInstance -ClassName Win32_Bios
      Get-Process | Select Id, ProcessName
    Get-Service | ? Status -match "Stopped" |Select
      DisplayName
  }
}
```

When this workflow is run, it is easy to see that these commands are run in parallel, as the output doesn't display in the order of execution. In fact, `CimInstance` completes first, then the services, and finally the processes and the exports show these were run in parallel. The order would change depending on what completes first. These workflows will be developed more in *Chapter 4, Windows Administration within VMware Administration*, when we mix VMware administration scripts with Windows administration scripts to fill out more remote development.

Quality control, consistency, and simplification

Although this is not the most exciting of sections, quality control, verification testing, and error handling are the most important areas of any script. Without testing and error handling, attempting to explain that the PoSH text in red is okay and the errors shown are normal, over and over, makes the perception of the script, for lack of a better word, bad. This section discusses the flow of the script, preparing what is expected and planning for what is not.

Revisiting documentation

Going back to the documentation is extremely important. Highlighting, as it is written, that the **#Gathers information to be used for building the VM later on** function is good but writing that Function X is for **#decision processing** and that the `VMName Record` is **#the name of the VM** is better.

Now, our function is given as follows:

```
<#This function gets a list of VMs and searches for the VMs with a V in
the name#>
Function Gather-VmsWithAV ($VMlist) {
  #Loop to place the entries in an Array
  Foreach ($a in $(get-VM $list | ? Name -Match "v")){
    $Output += $a.Name
    $Output += ","
    }
  #returns the Full array, Comma separated
  return($Output)
  }
#Call script to connect to vCenter
Invoke-expression -command ./connect-vCenter.ps1

#creates the full list of VMs from vCenter
$List = Get-vm

#Call the Function to process the names
Gather_VmsWithAV($list)
```

In comparison to the preceding code snippet, the difference is as follows:

```
Function A ($list) {
  Foreach ($a in $(get-VM $list | ? Name -Match "v")){
    $output += $a.Name
    $Output += ","
    }
  return($Output)
  }

Invoke-expression -command ./connect-vCenter.ps1
$List = Get-vm
A($list)
```

When the comments are in place, the code is readable and permits another individual to pick up the script and figure out what it is doing. Providing this as writing is extremely useful when there is more than one individual in the environment conducting the development. Consider variables as notes as well. If there is a variable named `$a` or a variable named `$largeInt`, these can be vague and hard to integrate if they are called much later in a long script. Attempt to use slightly more descriptive names, such as `$VMName`, `$vHost`, or `$VMList`, for common variables; it is easier to find and remember them.

The other point to be mentioned is the header of the script, where the input and output areas are placed. It assists in the understanding of what is supposed to be accomplished when these are highlighted before writing the script. It plans the script and assists in documenting the flow.

Script simplification

When writing the script, always consider reducing inputs from the user. A few years ago, as a leader of our local **VMUG** (**VMware User Group**) I developed a script for user registration. This script would get a list of users from VMUG HQ and would ask for the attendee's last name to search for. The script worked great until it came across people with the same last name. I had to ask the question, "If they searched for their last name and it found other people with the same last name, how do I return to the search results that have this consideration in mind?". I really had to think of how to rectify that simple issue when it changed the whole output's considerations. Don't assume that simple issues always have a simple method to fix. Sometimes, a complete rewrite of an area needs to occur to make the input/output feasible.

So considering the script example passed previously, the user list is an Excel spreadsheet that has **First Name**, **Last Name**, **E-mail Address**, **Company**, and **Title**. The user arrives at the meeting, is greeted by the registration attendant, and is asked to check in. The script must be able to read the spreadsheet to capture all the individuals that have pre-registered, as it needs to capture data for individuals that are in attendance but need to register right there and then, and an output spreadsheet needs to be created in order to be forwarded to HQ for processing. The following points are taken into consideration:

- **Input**:
    - Spreadsheet filename
    - Request for user information
    - Last name
    - Confirmation that the information is correct

  - Validation against predetermined information
  - If no valid comparison exists, prompt for complete user information

- **Output**:
  - Output spreadsheet filename
  - Display searched information

It all seems pretty simple until it is broken down into its individual components. The collecting the spreadsheet filename can be hardcoded and forced to retrieve a static name; however, do you validate the data that is being collected, or if there is a blank field that is vital to the data collection? Some of the techniques discussed earlier in the chapter, such as functions and using `invoke-expression`, can and will help with the processing and error handling of the data.

Error handling

Error handling is taking the guesswork out of selecting items. Point and click has become very popular, as it is much easier to select a radio button than to type in a text entry. How does someone validate the text to make sure that the entry is authentic and fits into the mold that is expected? For example, this may require ensuring the data that is being entered is a numeric value and not a special character, or even prevent users from accidently hitting *Enter* and making the field a Null or space. Creating functions that capture the entry and validate it is the most useful method to do. Let's attempt to create a sample function that will accept a range of numbers and validate the entry to that range:

- **Premise**: Create a function that receives a min and max in a range and validates the entry with a `True` or `False`
- **Inputs**: These are number values in a range
- **Outputs**: `True` or `False`
- **Assumptions**: It will be text-based only
- **Script**: Consider the following code snippet:

  `Test-FunctionInt.ps1`

```
<#
.Synopsis
Create a function that receives a min and max in a range and
validate the entry with a true or false
.Description
```

```
The Function will only accept numeric values, start of the range,
and end of the range. This script will also show the read of an
entry from a user and return a true if the number falls in the
range and a false if it doesn't.
.Input
Number
.Output
True or False
.Author
Chris Halverson
.Change Log
    11/6/2015
.FileName
    Test-FunctionInt.ps1
.Version
    Draft 0.1
#>

Function Test-NumRange {

Param($Min, $Max, $Num)
Process {
  $Num = [convert]::ToInt32($num)
If ($Num -ge $Min -and $Num -le $Max) {
    Return ($True)
}
  Else {
    Return ($False)
}
  }
}

Test-NumRange -Min 1 -Max 10 -Num $(Read-Host "Enter a number `
between 1 and 10?")
```

This is indeed a very simple example but it gets the point across of having a function that verifies a range will, no doubt, assist in the simplification of the input for the user. Adding a `write-host` function immediately following the `Test-NumRange` call that positions another request for input will help curb errors and receive accurate data for numeric data. What about string data?

- **Premise**: Create a function that receives a list of string values that needs an input tested against. The list comparison will return a `True` or `False` value
- **Inputs**: String based upon a predefined list of values
- **Outputs**: `True` or `False`
- **Assumptions**: None
- **Script**: Consider the following code snippet:

  `Test-FunctionStr.ps1`

```
<#
.Synopsis
Create a function that receives a list of values and validate the
input from Read-Host with a true or false
.Description
The Function will only accept String values that are defined in
$list. This script will repeat the input until the value that is
desired is inputted.
.Input
String based upon a list
.Output
True or False
.Author
Chris Halverson
.Change Log
    11/8/2015
.FileName
    Test-FunctionStr.ps1
.Version
    Draft 0.1
#>

Function Test-StrValue {

Param($List, $InAnswer)
```

```
Process {
      If ($list -contains $InAnswer) {
      Return($true)
      Break
        }
  Write-Host " Value not found, try again."
   Return ($false)
   }
}

Write-Host "Are you happy today?"
$List = "y", "n", "s"

Do {
    $Ans = Read-Host "[y]es, [n]o, [s]ort of?"
    $a = Test-StrValue -List $List -InAnswer $Ans
    }
    Until ($a -match $true)

Switch ($Ans) {
    List[0] {write-host "I am so glad to hear! "}
    List[1] {write-host "That is terrible, I hope you cheer
        up." }
    default {Write-Host "That is a none committal answer. `
       Maybe tomorrow."}
    }
```

Note that on one of the last lines of the script, there is a ` character that ends the line, and the line is continued on to the next line. This character tells the command interpreter that this line is not finished and there is more to come. It was used in this script to show the break in the line and also to help in copying word for word from the book. The margins in the book are of a certain length and occasionally the script line will break over to the next line, as was the case in this script. This is not always the case in the book but is for this example.

Verifying a "Yes/No/Maybe" scenario is fairly simple, but changing `List [n] {scriptblock}` could change the outcome dramatically. The function simplifies the outcome for the user and stops the script from being wrong by making an incorrect selection.

Verification testing

In the majority of IT projects, there is usually a testing phase that takes the final product and attempts to run it in a controlled manner so as to not affect the final production data. So, how would one test a script that is supposed to perform administrative work without taking down a production system? As this book is about VMware administration, a lot of testing can be performed through a virtualized framework:

- Cloning certain components or servers into a segregated environment
- Building a separate and segregated lab environment that the tests can run in

Using a script to build the clones and placing them in an appropriate `PortGroup` would be great to perform; however, in this instance, there is no way to perform this with the live system. So, running the clone through the web client or `VIclient` will need to be done in the only way it can be done, live, with no outages.

Create a virtual PortGroup using the following command lines:

```
$vHost = Get-VMHost | select -first 1
$vHost | New-VirtualSwitch -Name Internal
```

Once the required VMs are cloned, make sure that the network is using this new virtual switch and, because this is on one VMhost only, the machines will still be able to talk to each other but will not be able to talk to other machines beyond this virtual switch.

Version control and keeping previous versions

Version control, in this context, is only to make sure previous versions are kept as history and for reference going forward. As seen in all the previous scripts, there is a change log and a version number listed. This allows the practice of keeping a running tab of what is done in the script for developing and perfecting it. There is no reason why a script can't stop at a version 0.1 draft if it does what is required and does not need to be changed. The version number simply allows reader to understand what went into the stages of the script.

My personal preference is to mark the initial version as `0.1`. The first digit denotes a major version, whereas the second digit denotes a minor revision. Therefore, the 0 major version tells me it is the beta or draft of the script and the x.1 speaks to the first writing of the script. Usually, when I feel the script has been tested and I don't feel changes are needed, I relabel the script to version 1.0 or the final initial version. Then, if tweaks or changes are made, it moves to 1.1, 1.2, and so on.

What signifies a major version versus a minor version

In my case, if I change the script enough to alter a function or a small section of code, I will increase the minor version number. However, if I change how the script functions, such as going from a text-based gathering system to a graphical interface, or using a CSV instead of manually entering the data one entry at a time, I will increment the major version number. As this is my point of view, don't feel that this is set in stone; determine a standard for your team and stick to it.

Storing versions

For historical reasons, I recommend storing one previous major version and at least two minor versions. For example, if the current version is 2.3, the latest minor versions, 2.2 and 2.1, should be stored in a historical folder. This allows for easy referral, and if the new version doesn't work or has a bug it is easy to revert to a previous version.

Building a VM script using all the pieces

Consider the following parameters:

- **Premise**: Create a script that will prompt a user for information to build a new VM
- **Inputs**: This includes vCPUs, memory, datacenter, DRS cluster, valid datastore, Resource Pool, VM folder, customization specification, and data disk size
- **Outputs**: These are freshly built VM to specs
- **Assumptions**: Data gathering will be done on valid information from the virtual environment itself; no hard coded information will be valid unless specified in the script to the user (for example, the OS disk size)

- **Script**: Consider the following code snippet:

Connect-vCenter.ps1

```
<#
.Synopsis
    Does the lifting of connecting to a vCenter
.Description
    Starts VIMAutomation Module
    Gets Credentials
    Gets the vCenter name and connects to vCenter
.Input
    User Input
        Credentials
        vCenter Name
.Output
    None
.Author
    Chris Halverson
.Change Log
    11/6/2015
.FileName
    Connect-vCenter.ps1
.Version
    Draft 0.1
#>

#Call the Module
If ($(get-module -ListAvailable) -notmatch "VMware.VimAutomation.
Core") {
    Import-Module VMware.vimautomation.core
    }

#Test for stored credentials
If ($global:Cred -ne $null){
    write-Host -ForegroundColor Green 'Using the Credential
'$Cred.UserName
    }
```

```
Else {
    $global:Cred = Get-Credential
    }
[Console]::ForegroundColor = "Yellow"
$Global:VC = Read-Host "What is the name of the vCenter server?"

#Trying a ping of the vCenter Host
If (Test-Connection -ComputerName $Global:VC -Count 2 -quiet) {
    Connect-ViServer $Global:VC -Credential $Cred -ErrorAction
Inquire | Out-Null
    Write-Host "Connected to vCenter $VC " -ForegroundColor Green
    [Console]::Foregroundcolor = "Gray"
    }
Else {
    Write-Host "vCenter not found. Exiting script" `
   -ForegroundColor Red
    [console]::ForegroundColor = "Gray"
    Break
    }
```

Gather-Data.ps1

```
<#
.Synopsis
    Collects Information for build data
.Description
    Gets Appropriate data and outputs to a variable
.Input
    User Input
        VM Name
        Number of CPUs
        Amount of RAM in GB
.Output
    Variable
.Author
    Chris Halverson
```

```
.Change Log
    11/6/2015
.FileName
    Gather-Data.ps1
.Version
    Draft 0.2
#>

##Functions##
<#Get-Decision - simple function to ask for a y or n and return a
$true or $false#>
Function Get-Decision ($Decision){
  Switch ($Decision.toupper()){
    "Y" {$Decision = $True}
    "N" {$Decision = $False}
    default {"Enter a ''Y'' for a Yes or a ''N'' for No."
Get-Decision ($Decision)}
    }
  Return ($Decision)
  }

##Main Script

[console]::ForegroundColor = "Yellow"
$VMName = Read-Host "`nWhat is the name for the VM?"
if ($(Get-VM) -match $VMName){
  Write-Host -foregroundcolor Red `
'VM name already used. Exiting...'
  Exit
  }
[console]::ForegroundColor = "Green"

##Selection of DataCenter
$DC = Get-Datacenter
Switch ($DC.Count){
```

```
  "0" {Write-host -ForegroundColor Red 'No available Datacenters
exist! Exiting'
    Exit}
  "1" {$DCSel = $DC}
  default {
    Write-Host "`n"
    $menu = @{}
    for ($i=1;$i -le $DC.count; $i++) {
        Write-Host "$i. $($DC[$i-1])"
        $menu.Add($i,($DC[$i-1]))
        }
    Write-Host "`n"
    [console]::ForegroundColor = "Yellow"
    [int]$ans = Read-Host 'Choose the Datacenter'
    [console]::ForegroundColor = "Gray"
    $DCSel = $menu.item($ans)
    }
  }

##selection of DRS Cluster
$Cluster = $DCSel | Get-Cluster
Switch ($Cluster.Count){
  "0" {$ClusterSel = $Null}
  "1" {Write-Host -ForegroundColor Green 'One Cluster found,
Selecting' $Cluster.Name
    $ClusterSel = $Cluster}
  default {
    Write-Host "`n"
    $menu = @{}
    for ($i=1;$i -le $Cluster.count; $i++) {
        Write-Host "$i. $($Cluster[$i-1])"
        $menu.Add($i,($Cluster[$i-1]))
        }
    Write-Host "`n"
    [console]::ForegroundColor = "Yellow"
    [int]$ans = Read-Host 'Choose the Cluster'
```

```
    [console]::ForegroundColor = "Gray"
    $ClusterSel = $menu.item($ans)
    }
  }
$vHostSel = Get-Cluster $ClusterSel | Get-VMHost | Get-Random
Write-Host -ForegroundColor Green 'Selecting VMhost' $vHostSel.
Name

##selection of any resource pool
Write-Host -ForegroundColor Yellow 'Resource Pools are determined
as an arbitrary option and this script will not allocate the
Resource Pool. Manual move may be required.'

##This allows multitiered Folder structure to be shown
$FolderList = $DCSel | Get-Folder | ? {$_.Type -match "VM"} `
| ? {$_.Name -notmatch "Templates"}
Switch ($FolderList.Count){
  "0" {$FLSel = $Null}
  "1" {Write-Host -ForegroundColor Green `
'One Folder found, Selecting' $FolderList.Name
    $FLSel = $FolderList}
  default {
    Write-Host "`n"
    $menu = @{}
    for ($i=1;$i -le $FolderList.count; $i++) {
        Write-Host "$i. $($FolderList[$i-1])"
        $menu.Add($i,($FolderList[$i-1]))
        }
    Write-Host "`n"
    [console]::ForegroundColor = "Yellow"
    [int]$ans = Read-Host 'Choose the Folder'
    [console]::ForegroundColor = "Gray"
    $FLSel = $menu.item($ans)
    }
  }

##selection of a Template if any
```

```
$Tpl = $DC | Get-Template
Switch ($Tpl.Count){
  "0" {Write-Host -foregroundcolor Red `
'No Templates found! Unable to continue'
    Exit}
  "1" {Write-Host -ForegroundColor Green `
'One Template found, Selecting' $Tpl.Name
    $TplSel = $Tpl}
  default {
    Write-Host "`n"
    $menu = @{}
    for ($i=1;$i -le $Tpl.count; $i++) {
        Write-Host "$i. $($Tpl[$i-1])"
        $menu.Add($i,($Tpl[$i-1]))
        }
    Write-Host "`n"
    [console]::ForegroundColor = "Yellow"
    [int]$ans = Read-Host 'Choose the Template'
    [console]::ForegroundColor = "Gray"
    $TplSel = $menu.item($ans)
    }
  }

##selection of a customization specification
$CS = Get-OSCustomizationSpec
Switch ($CS.Count){
  "0" {Write-Host -ForegroundColor Red `
  'No Customization Specs found! Unable to continue'
    Exit}
  "1" {Write-Host -ForegroundColor Green `
'One Customization Spec found, Selecting' $CS.Name
    $CSSel = $CS}
  default {
    Write-Host "`n"
    $menu = @{}
    for ($i=1;$i -le $CS.count; $i++) {
```

```
        Write-Host "$i. $($CS[$i-1])"
        $menu.Add($i,($CS[$i-1]))
        }
    Write-Host "`n"
    [console]::ForegroundColor = "Yellow"
    [int]$ans = Read-Host 'Choose the CustomizationSpec'
    [console]::ForegroundColor = "Gray"
    $CSSel = $menu.item($ans)
    }
  }

##selection of Datastore
$DS = $vHostSel | Get-Datastore
Switch ($DS.Count){
  "0" {Write-Host -ForegroundColor Red `
'No Datastores found! Unable to continue'
    Exit}
  "1" {Write-Host -ForegroundColor Green `
'One Datastore found, Selecting' $DS.Name
    $DSSel = $DS}
  default {
    Write-Host "`n"
    $menu = @{}
    for ($i=1;$i -le $DS.count; $i++) {
        Write-Host "$i. $($DS[$i-1])"
        $menu.Add($i,($DS[$i-1]))
        }
    Write-Host "`n"
    [console]::ForegroundColor = "Yellow"
    [int]$ans = Read-Host 'Choose the Datastore'
    [console]::ForegroundColor = "Gray"
    $DSSel = $menu.item($ans)
    }
  }

##Selection of Network
```

```
$Net = $vHostSel | Get-virtualportgroup
Switch ($Net.Count){
  "0" {Write-Host -ForegroundColor Red `
'No Virtual PortGroup found! Unable to continue'
    Exit}
  "1" {Write-Host -ForegroundColor Green `
'One Virtual PortGroup found, Selecting' $Net.Name
    $NetSel = $Net}
  default {
    Write-Host "`n"
    $menu = @{}
    for ($i=1;$i -le $Net.count; $i++) {
        Write-Host "$i. $($Net[$i-1])"
        $menu.Add($i,($Net[$i-1]))
        }
    Write-Host "`n"
    [console]::ForegroundColor = "Yellow"
    [int]$ans = Read-Host 'Choose the PortGroup'
    [console]::ForegroundColor = "Gray"
    $NetSel = $menu.item($ans)
    }
  }

$Global:VMSpecs = @{
  Name = $VMName
  DC = $DCSel
  Cluster = $ClusterSel
  Host = $vHostSel
  Folder = $FLSel
  Template = $TplSel
  CustomSpec = $CSSel
  Datastore = $DSSel
  PortGroup = $NetSel
  }
```

Build-VM.ps1

```
<#
.Synopsis
    Builds a VM
.Description
    Calls the Connect-vCenter.ps1 Script to connect to vCenter
    Gathers Data from Gather-Data.ps1
    Builds VM based on Specs
.Input
    Called through other Scripts
.Output
    Builds VM
.Author
    Chris Halverson
.Change Log
    11/6/2015
.FileName
    Build-VM.ps1
.Version
    Draft 0.1
#>
#Resetting global variable
$Global:VMSpecs = $null

##Main Script##

##Connecting to the vCenter##
Invoke-Expression -command .\Connect-vCenter.ps1
Invoke-Expression -Command .\Gather-Data.ps1
If ($Global:VMSpecs -eq $null)
  {Exit}

##Build the VM
Write-Host -ForegroundColor Green 'Building VM...'

New-VM -Name $Global:VMSpecs.Name -vmhost $Global:VMSpecs.Host `
```

```
-Template $Global:VMspecs.Template -Datastore `
$Global:VMSpecs.Datastore -OSCustomizationSpec `
$Global:VMSpecs.CustomSpec | Out-Null

Get-VM -Name $global:VMSpecs.Name | Move-VM -Destination `
$Global:VMSpecs.Folder -Confirm:$false | Out-Null
Write-Host -ForegroundColor Magenta 'Powering up VM...'
Get-VM -Name $global:VMSpecs.Name | Start-VM -Confirm:$false | `
Out-Null
Write-Host -ForegroundColor Cyan 'Exiting Script'
```

Summary

This chapter summarized good coding practice, outlined the use of repeatable code, and ways to use that code for other general uses. It built a longer code base, provided methods for keeping the flow of development from concept, to thought, to action, and summed up a version control system to keep historical information into the script management structure, in order to help make the whole script traceable from inception to final iteration.

In the next chapter, we bring these VMware administrative scripts together and mash them with the standard Windows administrative script enabling the provisioning, operating, and decommissioning of a VM and establishing the Windows workflow as part of the same script.

4
Windows Administration within VMware Administration

Taking a step back, you realize that the rest of the department overheard the whole ordeal with the boss. Your coworker walks over and says "Don't worry, we are all working hard to make that so-called "office renovation" a big joke. The management knows you created the report and has been saving his job for years now. We have been asked to build a proper DevOps environment for the company with the $150k and you're likely going to be asked to lead it." Then you ask, "How do you know that?". The coworker quickly retorts, "I saw the pink slip on the boss's desk" he giggles. "It has been a long time coming."

Brinngg..., as you hear your desk phone ring, you pick up and say "Hello" followed by "Sure, Mr. Mitchells, I'll be right up." As the elevator moves up 10 floors to the CIO's floor, you reassure yourself that this is what the coworker was talking about. As you enter the giant corner office, you see the CIO and the CFO sitting and chatting on the couch discussing the latest sports game; they look up and greet you as you walk in, "We are glad you are here, and we have an interesting proposal for you."

The decision to build an Operations Development center, or "DevOps", needs to come from the top of an organization. Developing a script and an automation framework is a time saver in the long run but it requires capital and time to build up, not only in building human capital in terms of training and architectural understanding, but also in terms of time for the development of the automation. This chapter is going to deal with starting a DevOps practice, automating the VMware and GuestOS scripts in a single script, and introducing a little-known VMware application that, at the time of writing, is included in every purchased license of vCenter: vRealize Orchestrator.

- Building a DevOps practice

- Building a VMware PowerCLI Script that talks to the VMware infrastructure and Windows infrastructure
- Using a script host to control multiple machine scripts

What is DevOps?

When looking for a good definition for DevOps, I liked what Wikipedia had for the definition of the term:

> *"DevOps is a clipped compound of "development" and "operations". It is a culture, movement or a practice that emphasizes the collaboration and communication of both software developers and other information-technology (IT) professionals while automating the process of software delivery and infrastructure changes."*

Considering that this is such a new process, or movement, the techniques and expertise are limited. With that being said, there have been years of expertise gained through another term, named software delivery. Years ago, there were essentially two main providers of this type of tool and software delivery; Microsoft had their SMS server and Altiris also had an orchestration tool—both seemed to have limited practice to deploy applications and run scripts on the client's desktop for general management.

There was always trepidation when running these tools on servers due to the importance of the operations and management of the agents and the possibility of mistakenly running the wrong script on the exchange server for example. Batch (`.bat`) scripting was king at this point and VBScript was also used in a smaller capacity. Large companies and government bodies needed armies of system admins to control and operate these scripting methodologies. However, in the last few years, the number of servers has exploded and the budget to supply people to operate them has dwindled; therefore, the desire to automate has greatly increased.

Starting a DevOps practice

The consideration of starting a DevOps practice usually comes when the company has reached a certain size, number of VMs, or needs to reduce the number of administrators. It is a significant undertaking and typically requires additional time, capital, and the dedication to a platform or tool. Near the end of this chapter, we will go over a VMware tool named **vRealize Orchestrator** (previously known as vCenter Orchestrator), which provides workflow generation for building VMs, launching remote scripts, and providing a mechanism to run those PowerShell scripts with a common frontend. The best part about the tool is that if you have a vCenter license, Orchestrator (vRO) is free.

Starting a DevOps practice is similar to what was discussed in the previous chapter about **Software Control Management** (**SCM**), or keeping a software repository of the developed scripts. The tool needs to be chosen up front, because once the workflows and scripts are inputted the way the tool expects the inputs and outputs, it is difficult to switch (although not impossible).

DevOps needs to start with a personnel, who can program, who wants to program, and who is senior in the administration space. If the company has a development team, start with those individuals that have had previous infrastructure experience and are good at interfacing with the infrastructure team, as they are now the business to be supported. A typical structure that works is the pairing of resources, one from the junior side for developer and one from the senior for the administration.

vRealize Automation (another program from VMware) provides a good understanding of roles for this type of endeavor. In the following diagram, we see a single tenant structure and the roles associated. The graphic is taken from `http://pubs.vmware.com/vra-62/topic/com.vmware.ICbase/PDF/vrealize-automation-62-foundations-and-concepts.pdf` and explains a private cloud initiative.

Think of the tenant as the company's DevOps structure. Each business group would be a department within the IT organization (document management, business systems, and database team are some examples), whereas the Fabric groups would be similar to database systems, web systems, and app systems. Being part of operational transformation projects as a technical architect, this has been a key part of my role in many organizations. This diagram definitely is larger than a typical DevOps framework, but provides some context for where DevOps may fit in:

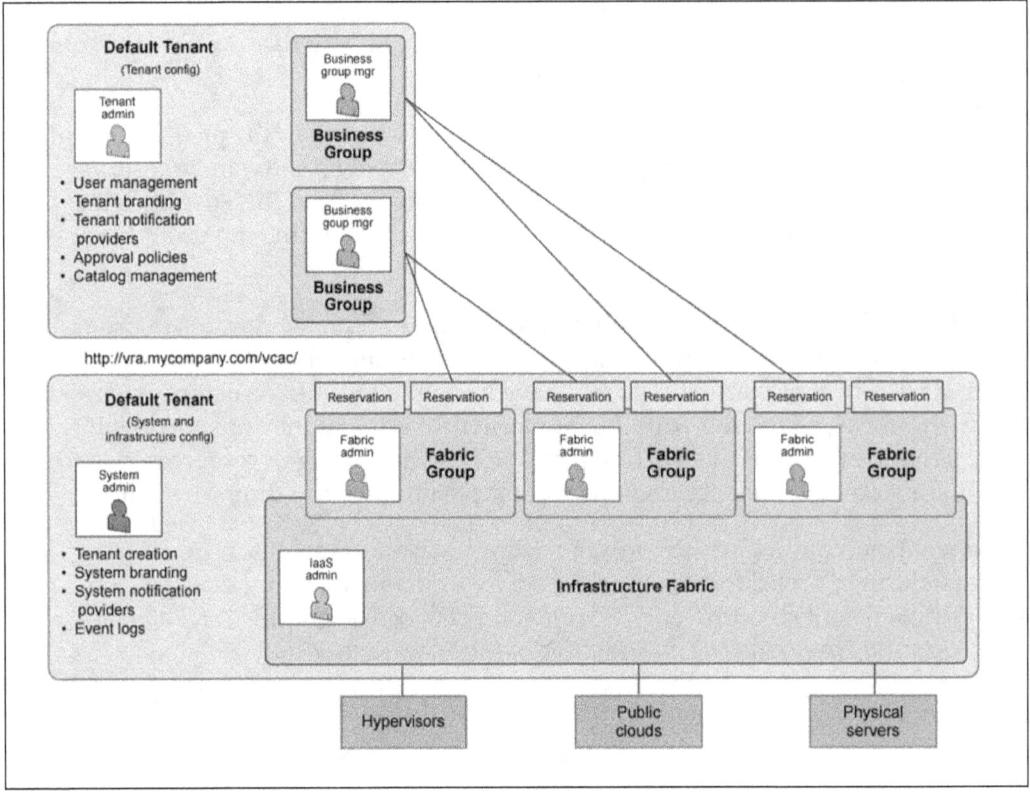

The roles seen here include **Tenant Admin**, **System Admin**, **IaaS Admin**, **Fabric Admin**, and **Business Group Admin**. Each role helps to define ownership or responsibility of an area of the whole solution. DevOps creates a lifecycle for how to manage the creation, operation, and decommissioning of a server. It simplifies the entire server administration to enable quicker responses, automated alerting, rectification, and a means to help reduce the amount of administrators.

Defining the roles

The following table contains example roles that can be designed as part of a structure, keeping in mind that these don't need to have an individual per role and can have multiple personnel per role depending on the size of the DevOps requirement:

Role	Responsibility	How it's assigned
Administrator	Manage identity stores by enabling access through user and group provisioning. Management of the DevOps framework from the script host structure to the automation tool. Responsible for ensuring sizing for the environment and keeping the cycle of commissioning and decommissioning within the bounds of the physical limitations of the hardware.	Apart from having a key understanding of the framework, this individual helps ensure the tools are ready to provide timely updates to the system and reporting on usage.
Service Architect	When a new script or workflow is needed, the service architect provides the why and how for the script. Does the effort equal the outcome? Determines repeatability of the script. How often will the service be needed and how frequently will it be needed to run? Decides on the complexity of the script. If the script or workflow has many different parts and orchestration is critical, is it better to be automated or manual? Assigns risk to the running of the script or workflow to a standard change management process.	Mainly requires an understanding of administration and the ability to read workflows.

Role	Responsibility	How it's assigned
Automation Engineer	Automation and development experience is needed for this role. Once the service architect has determined the need, the automation engineer will automate it.	This person should possess strong development ideology and an understanding of infrastructure administration. This role should involve two individuals, one would be a junior to the intermediate developer and the other would be a senior to the expert level infrastructure administrator. The idea of using two resources allows for two points of view to be factored into the final product.
Quality Assurance (QA) and Testing	QA can be included in the role responsibility of the automation engineer, but it needs to be someone who has not worked on the original development.	Someone who understands how to automate but did not develop the workflow or script.
Support User	The support user will provide support to users of the workflows and scripts. The tasks done are documenting the errors that occur and controlling the bug tracking of the scripts.	Helpdesk or service desk individuals can provide this service.
Approval Administrator	This is a responsibility track and can fall into the hands of the QA individual. This role will move the script from development, to testing, to production and assign an owner to the script or workflow	This is the QA engineer, or the automation engineer that developed the script or workflow
Approver and Owner	This is a simple role that approves the running of the script or workflow; It checks whether there are risks that are high enough or else if the defining version has too many bugs then it decides to return it to the development/test stage.	This is the manager of the DevOps group

An example process

Mr. Mitchells enables you to have the DevOps department and asks for a specific report to be produced on a weekly basis for the auditors. The report has a number of requirements due to a compliance need, and the application group size is 200 VMs in the farm. The application farm needs to have their times synchronized to a central NTP, server and this is done every hour. The servers have an NTPSync service that does this and it must be running at the top of the hour, every hour, and if for whatever reason the service is stopped, it must be started and the application owner must be alerted that it was stopped.

The next requirement is for the Windows OS disk space; the application on occasion balloons a file by 10 GB to help accommodate a temporary dump file of the application on 10 of the 200 VMs. 20% of the `C:` drive must be available for this occurrence at any point in time. Finally, the last requirement is that all 200 VMs must reside on a single NFS datastore and that the datastore must be able to have enough room for the VMs that need to be on it. If the datastore is changed on a VM for any reason, the machine must automatically be returned through Storage vMotion.

Some of the requirements have now been documented and the service architect begins by documenting and validating them in a script process.

Requirement ID	Requirement	Notes
R01	200 VMs have to have service monitored and restarted on an hourly basis	NTPSync Service
R02	OS drives must have 20% or 10 GB free at all times (named 10 VMs)	
R03	All VMs must be in an acceptable datastore	

After building the requirements table, the architect needs to validate whether the service is real, whether it can be restarted, and what are the consequences if the service isn't started on an hourly basis. If the 10 GB free space requirement is based on a random 10 VMs or if they are specifically named instances, a listing of the consequences is prepared when the 10 GB available space barrier isn't met. Lastly, the architect also needs to validate whether all 200 VMs are running on one datastore, whether there is another reason they must be maintained with that requirement, and whether there are going to be any constraints, such as limited connectivity to all the nodes or a situation in which one node is the controller of many worker nodes.

Once confirmed, the gathering of requirements leads to the design of the script. This work would be conducted with the Automation Engineers to build the workflow. What is the most stringent requirement? Is there a least restrictive requirement? Once confirmed and designed, the Automation Engineers create the script. Once running and functional, it is moved to the QA and testing area. Once they run through some testing scenarios and the running of the script/workflow, it is sent for approval from the Approval Admin and then deployed.

Day 2 operations (support users) would then be responsible for any bug tracking and operational issue reporting with the workflow. This is handed back to the Automation Engineers to resolve the issues. The approver grants access to the script and the Administrator grants access to the workflow based on that approval.

Duties of the Automation Engineer

Based upon the preceding project, this book ideally tries to cover the Automation Engineer's task list, how to adapt the requirements gathering from the Service Architect, and how to script them. This section outlines the script process for building a VMware PowerCLI-based script, combining it with the Windows-based host PoSH script, and returning output.

Building the script

During the process detailed previously, there were three specific requirements:

- 200 VMs in a common VM-based folder
- All servers must have the common NTPSync service running and be tested once per hour
- All servers must have 10 GB of free space on the `C:` drive
- All VMs must be in a common datastore

There is a need to determine whether the scripts can be run independently or must be part of the same script. There is no requirement to have them run together, so we will start with independent scripts for each requirement.

Mixing PowerCLI and standard PowerShell

The idea of running PowerCLI within an environment is not new. As shown in earlier chapters, PowerCLI has been around for some time and there has always been a desire to do complete end-to-end cycle. Many one-line scripts, which I personally have seen, generally deal with a single purpose, a single task, and a means for saving a couple of clicks. These are great for simple processes but as the previous chapter introduced, automation saves time and removes the need for repetitive tasks.

What about spanning multiple systems? Perform a script launch, discover it is a Linux machine, and send a BASH script to change the IP on the fly; edit a `resolv.conf` file; or perhaps creating a build script that would generally be reserved for an expensive and complicated configurator program, which installs and configures patches and certain programs; there are numerous possibilities. The next section introduces these concepts.

Building a Windows script host

This sounds complicated but, truly, it isn't. *Chapter 3, Enhancing the Scripting Experience*, discusses working on `WinRM quickconfig`, and allowing scripts to be run from other machines. That is all a script host is—an alternate computer that is configured to launch and run a number of PowerShell processes from an orchestrator or administrative workstation. The script host utilizes its processing power without affecting the administrator's launch point.

What are the requirements? A little forethought should exist in the choosing of a script host:

- PowerShell needs processing, especially if multiple scripts will be running simultaneously on a single machine. Think about preparing a virtual machine that will be used for running the scripts—a Windows 2012 R2 Server configured in **server core** mode so other envious System Admins will not be tempted to install that "utility" application.

> **Windows 2012 R2 and server core**
>
> Windows server core is capable of having the GUI portions added to it, and the method to do it is found on many websites on the Internet. Although this can occur, generally, more work than most Admins want to tackle, due to the simplicity of deploying another VM for their purposes.

- PowerShell also uses the processor and RAM, so don't be tempted to under-provision it; provide more than the typical, 1 CPU and 2 GB of RAM. Consider how many synchronous scripts maybe running and size it accordingly. (Start with two CPUs and 4 GB of RAM and allow ramping up using hot-add CPU and memory.)
- Install the VMtools by using **Install/Upgrade VMware Tools** in the VIClient or **Install Tools** in the web client. The prompt will typically not come up on a server core install, so the use of command line is required. server core defaults to a `cmd.exe` administrative prompt at startup and launching a PowerShell instance is as simple as typing `PowerShell`.
- Run the `WinRM quickconfig` command to enable WinRM and remote PoSH capabilities. If the system was installed using the tried and true method of installing via the CD or ISO file, the computer will be the typical `WIN-LIGJEI224JFN` type name. To rename the computer, type `Rename-Computer ComputerName ; restart-computer`; once the computer is rebooted, the next step is to add the computer to the appropriate domain by typing `Add-Computer` and the cmdlet prompts for credentials and the domain name.

Using commands to enable remote desktops on a Windows 2012 R2 server

Two commands enable remote desktops on the server core instance. One method is to change the registry entry to deny all remote sessions and the other is to allow the remote sessions through the firewall. If Group Policy is enabled in the environment to turn these options off, then you are ahead of the game. If not, type these two commands to enable access to the desktop:

```
Set-ItemProperty -Path 'HKLM:\SYSTEM\CurrentControlSet\Control\Terminal
Server' -Name "fDenyTSConnections" -Value 0


Enable-NetFireWallRule -DisplayGroup "Remote Desktop"
```

Once these two commands are run, RDP is enabled on the server (yes, this even works on GUI-based servers as well.) Here are some other setup ideas:

- Install PowerCLI:

  ```
  .\VMware-PowerCLI-6.0.0-3205540.exe
  ```

- Add the Active Directory modules:

  ```
  Add-WindowsFeature RSAT-AD-PowerShell
  ```

- Map a central repository store and make it persistent:

```
New-PSDrive -Name X -Root \\NAS_Server\Scripts -Persist
-PSProvider FileSystem
```

Take note that a base install of server core, by default, doesn't import all the modules for use. The PoSH process, if all the modules are imported, will balloon the memory usage. When testing this, I personally ran `Get-Module -ListAvailable | Import-Module` and the memory usage went from 41 MB to 329 MB. Just loading the VMware PowerCLI modules boosted the usage to 80 MB, so consider this when running remote scripts; the offload does increase usage on the script host.

Running PowerCLI and Windows Administration

How does someone build a script that will create all the VMware components and the Windows functions? There are numerous ways to do this: one is by building separate scripts on every VM that a central script can call and another is to use a script that can run remotely from a central script host.

When considering VMware Administration, as seen in previous chapters, the vCenter or ESXi host must be connected before any information can be received. Windows administration using a script host has a similar requirement. `New-PSSession -ComputerName` attempts to connect to the designated script host to run the script. A major concern that needs to be addressed is connectivity; making sure the host and guest have the same ability to see each other is key to remote administration.

WinRM uses TCP port `5985` to communicate from the remote scripting host to the client or target of the script. As part of the `WinRM quickconfig` command, it opens the local host firewall to accept the connection, but this is not the only part of the requirement. Network-based firewalls, segregated nonroutable subnets, and client-based firewalls will prevent the ability to manage and control the clients.

A script example

Based on the requirements built in the previous section of the example process, a listing of each requirement is shown as follows. This example takes the requirements and begins the script development process. Starting with requirement R03, it shows you *how to check on VMs that are in a specific folder* and it is best to start with some sample code. The following commands can get the datastore, ensure the VMs are in the right datastore, and move the VMs if they are not. Thankfully, this doesn't have to occur more than once a day.

Run `.\connect-vCenter.ps1` to connect to the vCenter from the previous chapter:

```
Get-Folder <foldername> | Get-VM
$(Get-Folder <foldername> | Get-VM).Count
Get-Folder <foldername> | Get-VM | Get-Datastore | % Name
Get-Folder <foldername> | Get-VM | Move-VM -datastore <Datastore>
```

The following commands are meant to be individual commands that will help for each of the functions to be conducted. These are not complete by any means, but planning partial commands does help in the design of the final script:

- **Requirement R01**: This makes sure each of the servers in the list has the service running, and starts it if it is not. This one has to be scheduled once every hour and could be problematic if there is an issue with the service itself; therefore, a reporting feature should be added to the script to send an e-mail or alert if the service doesn't start:

  ```
  Get-Service | ? Name -match "NTPSync"
  If ($(Get-Service | ? Name -match "NTPSync").Status -ne "Running")`
  {Set-Service -name "NTPSync" -erroraction
  ```

- **Requirement R02**: This requirement is to check each VM frequently to see whether there is 10 GB free on all of the OS disks. Conceptually, this one is a lot trickier as the script needs to run almost continually, run solely on the guest, or examine and establish a predetermined time check for a daily or hourly time slot:

  ```
  If ([system.math]::round((get-psdrive -Name C | % Free) / 1GB, 2)
  -lt 10) {<Script Block>}
  ```

  Thankfully, `Get-PSDrive` produces the majority of the information needed, but it needs a script within the `If` statement that can produce the reporting functionality. So, we now need to determine how the reporting will be done.

- **Reporting**: Since the original request was an e-mail report to the manager, the output could be an e-mail so that the entire script would run and automatically e-mail the manager. However, from what I have found in the past, sanitizing the data before sending it in a report to management is critical. You need to check whether the report has appropriate information by fully vetting it before sending; if it has limited scope or if there is a problem, then sending an e-mail would be sufficient. Ideally, it should contain some sort of approval process where the typical report would go to the admin, certain conditions are examined, and this would allow the administrator to rectify the issue before the final alert e-mail to the manager:

  ```
  Send-MailMessage -to "Mr. Mitchell (Mr.Mitchell@company.com)"
  -from "Scripting (no-reply@company.com)" -subject "Disk Size Error
  Report" -attachments "Error.csv" -smtpserver smtp.company.com
  ```

Thankfully, PowerShell has a `Send-MailMessage` cmdlet and it works well for this requirement.

Thinking through the script

After getting some valid commands that can be run to get the approximate information, the next step is to think through how to do this. Firstly, there is a need to check each server for the disk space requirement, as we saw before, either continually or on a timed schedule. Performing a check every second will likely expend resources that are unnecessary; the risk of a drive filling up in that time schedule is unlikely. Checking continually will expend too much resources and the risk of a drive filling up in seconds is unlikely. Using a scheduled task that runs once every 30 minutes will be more than sufficient for typical workloads.

Always examine the change rate on a VM before coming to this conclusion. If the typical change rate is much higher and the risk that it will fill up in less than 30 minutes is significant, adjust the frequency accordingly.

Next, confirm whether the scripts will run on the VMs themselves or on a script host. If the script host is used to launch the local scripts, this method can be used:

```
$Cred = get-credentials
$PS = New-PSSession -ComputerName VM1 -Credential $Cred
Enter-PSSession $PS
$output = [System.Math]::Round((Get-PsDrive -Name C | % Free)/1GB, 2)
Exit-PSSession
```

The issue with this is that `$output` is stored in the PowerShell session on the remote computer and is deleted once the `PSSession` is disconnected. So, in the preceding example, the `$output` variable still exists but is not accessible by the script host. `Invoke-Command` allows the capturing of data through the cmdlet and in this case should be used:

```
$Cred = get-credentials
$PS = New-PSSession -ComputerName VM1 -Credential $Cred
$Output = Invoke-Command -Session $PS -ScriptBlock `
{[System.Math]::Round((Get-PsDrive -Name C | % Free)/1GB, 2)}
```

So the question is whether to run each command in a single command set through this type of command, or to create a script that the `Invoke-Command` would call, run, and capture? Writing a command to call a script on the running guest is simple enough, but what is the best way to push the `.ps1` file to the remote machine? Or is embedding the script in the script block the ideal way to run it? Will the output be brought back to the host system or leave it on the guest and output to a shared text file for parsing later?

Considering that 200 machines have been checked, the original 200-VM list needs to be input through a text file or hardcoded in the script. This can be done in a .csv file, but an easier way is to parse the DNS names from vCenter. So after all this thought, we are building this as `Get-Folder` in vCenter, pulling each of the VMs, finding the DNS name from each VM in the folder, checking for connectivity to the VM, running `Invoke-Command` for the ones that reply, outputting to a hash table, and eventually sending the whole report via e-mail for the desired time frame. Easy, huh?

Building the script

We'll consider the following points when building the script:

- **Premise**: Get a list of 200 VMs in a folder and perform a Windows service check and a free disk check
- **Inputs**: The folder to be examined for Windows script, the e-mail account to send to, and so on
- **Output**: Report containing violations of the requirements sent by an e-mail
- **Assumption**: The account to connect to vCenter has local administrator rights to the Windows VMs to be checked
- **Script**: Consider the following script:

  ```
  Report-AppSettings.ps1
  <#
  .Synopsis
  ```

```
  Get list from folder in vCenter, produce report of
  violations to the rules.
.Description
  Connects to vCenter using previously written
  Connect-vCenter.ps1 and requests Folder to scan
  Synchronously gets VMs and connects to each one prompting
  for disk size, and if the NTPSync Service is running.
   Reports on deviations and starts service if stopped.
.Input
   Credentials through Connect-vCenter.ps1 switch
  vCenter Name
  vCenter folder name
  Email account to send report
.Output
  Report
.Author
  Chris Halverson
.Change Log
  12/16/2015
.FileName
  Report-AppSettings.ps1
.Version
  Draft 0.1
#>
### Define the command to email in case of a failure.
$CmdScript = "Send-MailMessage -to Mr.Mitchell@company.com `
-from no-reply@company.com -subject "Disk Size Error Report" `
-Body $OutString -smtpserver smtp.company.com"

# Define the Service $Svc Variable to what you are looking for.
$SVC = "NTPSync"

#Invoke connect-vCenter.ps1
Invoke-Expression -command .\Connect-vCenter.ps1

<#
```

```
To get the folder to be selected the script will use a selection
mechanism that shows the available Datacenters first. Once
the Datacenter is selected, then the folders can be shown and
selected.
#>

$DC = Get-DataCenter
Switch ($DC.Count)
{
  "0" {
    Write-host -ForegroundColor Red 'No available
    Datacenters exist! Exiting'
    Exit
  }
  "1" {$DCSel = $DC}
  default {
    Write-Host "`n"
    $menu = @{}
    for ($i=1;$i -le $DC.count; $i++)
    {
      Write-Host "$i. $($DC[$i-1])"
      $menu.Add($i,($DC[$i-1]))
    }
    Write-Host "`n"
    [console]::ForegroundColor = "Yellow"
    [int]$ans = Read-Host 'Choose the Datacenter '
    [console]::ForegroundColor = "Gray"
    $DCSel = $menu.item($ans)
  }
}
$Menu = $Null
$Ans = $Null

$Folder = Get-DataCenter | Where Name -match $DCSel | Get-Folder `
-type VM | Where Name -NotMatch "vm"
Write-Host "`n"
$Menu = @{}
```

```
for ($i=1;$i -le $Folder.Count;$i++)
{
  Write-Host "$i. $($Folder[$i-1])"
  $Menu.Add($i,($Folder[$i-1]))
}
Write-Host "`n"
[console]::ForegroundColor = "Yellow"
[int]$Ans = Read-Host 'Choose the Folder to Validate against'
Write-Host "`n"
[console]::ForegroundColor = "Gray"
$FolderSel = $Menu.item($Ans)


<#
Folder is selected, setting a pre-determined datastore for the
VMs to be placed in.
#>
$DataStore = Get-DataStore SYN-iSCSI-SSD


## Pull VM data from the selected folder.


$VMs = Get-Folder -Name $FolderSel | Get-VM


Write-Host -foregroundColor Cyan "Confirmed VM count is
$($VMs.Count), beginning parsing data"


## Begin Loop per VM


Foreach ($VM in $VMs)
{
  Write-Host "`nScanning $VM "
  Write-Host -foregroundcolor Green `
  "Testing the Power state of the VM"
  If ($VM.PowerState -match "Off") {
    Write-Host -foregroundcolor Red `
    "Machine is not Powered On, bypassing VM"
    $OutScript = "$($VM.Name) Not Powered On"
```

```
    Invoke-Expression $CmdScript
  }
  Else {
    Write-Host -ForegroundColor Yellow `
    'Examining Datastore Location'
    Write-Host "The DataStore - $($VM | Get-Datastore) - is
    currently being used"
    If ($($VM | Get-Datastore) -ne $DataStore)
    {
      $VM | Move-VM -Datastore $DataStore -Confirm:$False |
      out-null
    }
    $view = $VM | Get-View

    ##Verify the DNS name from VMtools

    $VMName = $View.Summary.Guest.HostName
    Write-Host -ForegroundColor Green "`tGetting DNS Name ...
    $($VMName)"
    ##Test Connectivity to the VM through IP connectivity

    If (Test-Connection -ComputerName $VMName -Count 2)
    {
      ##Launch Remote PowerShell Session
      $PS = New-PSSession -Computername $VMName -Credential
      $Cred
      ##Check the DiskStatus
      $DiskOut = Invoke-Command -Session $PS -ScriptBlock `
      {[System.Math]::Round((Get-PsDrive -Name C | `
      %Free)/1GB, 2)}
      Write-Host -foregroundcolor Yellow "`tDisk Size of
      GuestOS is $DiskOut GB"
      ##Check the Service Status
      $SvcStatus = Invoke-Command -Session $PS -ScriptBlock {
        Get-Service | ? Name -match $SVC
      }
      Enter-PSSession $PS
```

```
        ##Has to redefine the SVC variable due to this running
        inside of a PSsession
        $SVC = "NTPSync"
        If ($(Get-Service | ? Name -match $SVC).Status -ne
        "Running")
        {
          Start-Service -name $SVC -erroraction SilentlyContinue
        }
        Exit-PSSession
        $SvcStatus = Invoke-Command -Session $PS -ScriptBlock {
          Get-Service
        }
        $SvcStatus | ? Name -eq $SVC | FT -HideTableHeaders
      }
    }
  }
```

```
Output:

PS C:\Users\chrish.TestLab\Scripts> .\Report-AppSettings.ps1
Using the Credential chrish@TestLab
What is the name of the vCenter server? : vcsa
Connected to vCenter vcsa

1. TestApplication
2. NetworkDevices
3. FileServices
4. Desktop
5. AD

Choose the Folder to Validate against : 5

Confirmed VM count is 2, beginning parsing data

Scanning vDC2
Testing the Power state of the VM
```

```
Examining Datastore Location
The DataStore - SYN-i-SSD - is currently being used
  Getting DNS Name ... vDC2.TestLab.Local
  Disk Size of GuestOS is 32.44 GB

Running NTPSync                    NTP Sync          vDC2.TestLab.
local

Scanning vDC1
Testing the Power state of the VM
Examining Datastore Location
The DataStore - SYN-i-SSD - is currently being used
  Getting DNS Name ... vDC1.TestLab.local
  Disk Size of GuestOS is 35.86 GB

Running NTPSync                    NTP Sync          vDC1.TestLab.
local
```

The script will send an e-mail to Mr. Mitchells if the VM is turned off. There is more possibility to that output expanding what other automated report can occur, as well as changing the script to run as a job, use arguments to enable batch processing, and setting it to run at given points in time. The outcome here highlights that a PowerCLI script can be run to collect VMs in a folder and run Windows-based cmdlets against them remotely.

Invoke-VMScript configuration

A simple example of this cmdlet follows the same thread as the last script example of getting the services on the VM:

```
Invoke-VMScript -vm $(Get-VM VM01) -ScriptText "Get-Service"
```

The resulting output is from the VM in question encompassed in the | (pipe) and - (dash or hyphen) symbols, as seen in the following screenshot. The output shows what the VM would display if it were run locally, as follows:

```
$out = Invoke-VMScript -vm $(Get-VM VM01) -ScriptText "Get-Service"
```

The variable captures the data, as seen in the following screenshot:

```
ScriptOutput
------------


    Directory: C:\Windows\system32\WindowsPowerShell\v1.0\Modules


ModuleType Name                               ExportedCommands
---------- ----                               ----------------
Manifest   ADRMS                              {Update-ADRMS, Uninstall-ADRM...
Manifest   AppLocker                          {Set-AppLockerPolicy, Get-App...
Manifest   BestPractices                      {Get-BpaModel, Invoke-BpaMode...
Manifest   BitsTransfer                       {Add-BitsFile, Remove-BitsTra...
Manifest   CimCmdlets                         {Get-CimAssociatedInstance, G...
Script     ISE                                {New-IseSnippet, Import-IseSn...
Manifest   Microsoft.PowerShell.Diagnostics   {Get-WinEvent, Get-Counter, I...
Manifest   Microsoft.PowerShell.Host          {Start-Transcript, Stop-Trans...
Manifest   Microsoft.PowerShell.Management    {Add-Content, Clear-Content, ...
Manifest   Microsoft.PowerShell.Security      {Get-Acl, Set-Acl, Get-PfxCer...
Manifest   Microsoft.PowerShell.Utility       {Format-List, Format-Custom, ...
Manifest   Microsoft.WSMan.Management         {Disable-WSManCredSSP, Enable...
Script     PSDiagnostics                      {Disable-PSTrace, Disable-PSW...
Binary     PSScheduledJob                     {New-JobTrigger, Add-JobTrigg...
Manifest   PSWorkflow                         {New-PSWorkflowExecutionOptio...
Manifest   PSWorkflowUtility                  Invoke-AsWorkflow
Manifest   ServerManager                      {Get-WindowsFeature, Add-Wind...
Manifest   TroubleshootingPack                {Get-TroubleshootingPack, Inv...


    Directory: C:\Program Files (x86)\VMware\Infrastructure\vSphere
    PowerCLI\Modules


ModuleType Name                               ExportedCommands
---------- ----                               ----------------
Binary     VMware.VimAutomation.Cis.Core
Binary     VMware.VimAutomation.Cloud
Manifest   VMware.VimAutomation.Core
Binary     VMware.VimAutomation.HA
Binary     VMware.VimAutomation.License
Binary     VMware.VimAutomation.PCloud
Manifest   VMware.VimAutomation.SDK
Binary     VMware.VimAutomation.Storage
Binary     VMware.VimAutomation.Vds
Binary     VMware.VimAutomation.vROps
Binary     VMware.VumAutomation
```

The output is a single string and will need some massaging to pull any specific information out of the `BlockText`. Rewriting the original command with a split can be used to divide the output into an array. Unfortunately, this string manipulation requires `ScriptOutput` to first be processed in one line, and then the string to be changed on the second line:

```
$out = $(Invoke-VMScript -vm $(Get-VM VM01) -ScriptText `
"Get-Module -listAvailable").ScriptOutput
$out = $out.split("`n`r")
```

However, the `$out` variable has blank lines for the first couple of items in the array. So, some trickery with string manipulation is in order:

```
$out = $(Invoke-VMScript -vm $(Get-VM VM01) -ScriptText `
"Get-Module -listAvailable").ScriptOutput
$out = $out.split("`n`r", [System.StringSplitOptions]::RemoveEmptyEntri
es)
```

This gets rid of the empty lines and compacts the data:

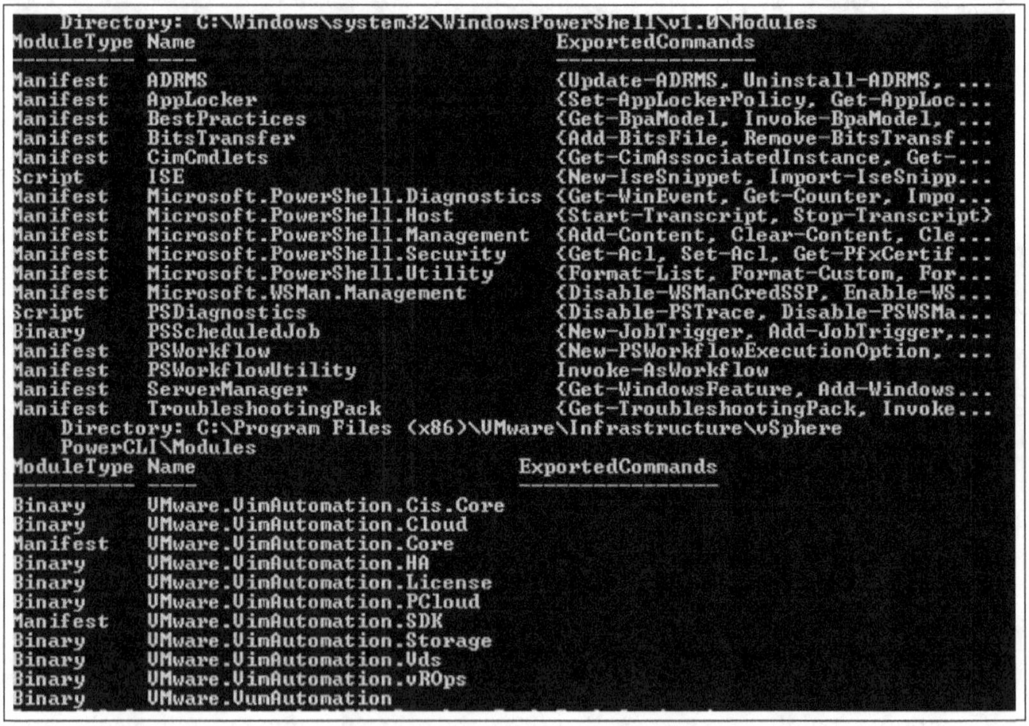

```
    Directory: C:\Windows\system32\WindowsPowerShell\v1.0\Modules
ModuleType Name                                ExportedCommands
---------- ----                                ----------------
Manifest   ADRMS                               {Update-ADRMS, Uninstall-ADRMS, ...
Manifest   AppLocker                           {Set-AppLockerPolicy, Get-AppLoc...
Manifest   BestPractices                       {Get-BpaModel, Invoke-BpaModel, ...
Manifest   BitsTransfer                        {Add-BitsFile, Remove-BitsTransf...
Manifest   CimCmdlets                          {Get-CimAssociatedInstance, Get-...
Script     ISE                                 {New-IseSnippet, Import-IseSnipp...
Manifest   Microsoft.PowerShell.Diagnostics    {Get-WinEvent, Get-Counter, Impo...
Manifest   Microsoft.PowerShell.Host           {Start-Transcript, Stop-Transcript}
Manifest   Microsoft.PowerShell.Management     {Add-Content, Clear-Content, Cle...
Manifest   Microsoft.PowerShell.Security       {Get-Acl, Set-Acl, Get-PfxCertif...
Manifest   Microsoft.PowerShell.Utility        {Format-List, Format-Custom, For...
Manifest   Microsoft.WSMan.Management          {Disable-WSManCredSSP, Enable-WS...
Script     PSDiagnostics                       {Disable-PSTrace, Disable-PSWSMa...
Binary     PSScheduledJob                      {New-JobTrigger, Add-JobTrigger,...
Manifest   PSWorkflow                          {New-PSWorkflowExecutionOption, ...
Manifest   PSWorkflowUtility                   Invoke-AsWorkflow
Manifest   ServerManager                       {Get-WindowsFeature, Add-Windows...
Manifest   TroubleshootingPack                 {Get-TroubleshootingPack, Invoke...
    Directory: C:\Program Files (x86)\VMware\Infrastructure\vSphere
    PowerCLI\Modules
ModuleType Name                             ExportedCommands
---------- ----                             ----------------
Binary     VMware.VimAutomation.Cis.Core
Binary     VMware.VimAutomation.Cloud
Manifest   VMware.VimAutomation.Core
Binary     VMware.VimAutomation.HA
Binary     VMware.VimAutomation.License
Binary     VMware.VimAutomation.PCloud
Manifest   VMware.VimAutomation.SDK
Binary     VMware.VimAutomation.Storage
Binary     VMware.VimAutomation.Vds
Binary     VMware.VimAutomation.vROps
Binary     VMware.VumAutomation
```

And now, there can be a quick search for a key piece of data, for example, if the `VimAutomation` module is available:

```
If ($out -match "VimAutomation.Core") {$True}
```

Running remote on a Linux guest

Attempting to cross platforms is always difficult and this type of crossing is also hard in PowerShell. Considering that PoSH does not run on Linux (for obvious reasons), there has to be a medium that pushes alternate programming languages to the VM.

This has to be done through one of two methods. The first method is `Invoke-VMScript`, in which a PowerCLI command talks to the VMTools of the VM and launches a script block that the VM can understand through the VMTools. It is much harder than it would appear on the surface. The second method is by launching an SSH connection and running the command as if the administrator is plunking away on the keyboard. It is a little easier to support and is more secure than the first method.

Basic BASH

A basic BASH script consists of components similar to a batch or PowerShell script, using a definition of the language and a list of commands in order to perform the function. Typical Linux BASH is used for navigating around the shell, running file operations, or running scheduled tasks (cron).

A simple example can consist of a file listing, using the command `ls -la`, or checking the disk utilization with `df -h`, and these can be run within the `Invoke-VMscript` cmdlet. `ScriptOutput` is exactly the same as the Windows example previously and it is just as simple to pull the screen output.

`Invoke-VMscript` is capable of working with PowerShell and Batch for Windows and BASH for Linux machines. By default, the cmdlet looks at the GuestOS and determines the default language to use. PowerShell is the default on Windows and BASH is the default on Linux.

Invoke-VMScript configuration

The first thing to do with a Linux connection is to set the credentials; unless Active Directory authentication is utilized, setting the credentials with `$linuxCred = Get-Credentials` will be needed beforehand:

```
$linuxCred = Get-Credentials
```

This brings up the typical prompt, as follows:

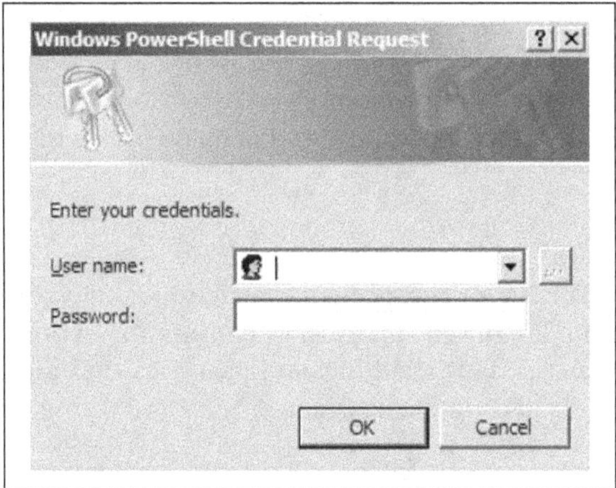

Enter `root` or the user account to access the Linux GuestOS and the password. This will be used for the `Invoke-VMScript` cmdlet next:

```
$(Invoke-VMScript -vm Linux01 -guestCredential $linuxCred -ScriptText "df
-h").ScriptOutput
```

This will produce this type of output:

```
Filesystem                              Size  Used Avail Use% Mounted on
/dev/sda3                                11G  3.7G  6.5G  37% /
udev                                    4.0G  168K  4.0G   1% /dev
tmpfs                                   4.0G   40K  4.0G   1% /dev/shm
/dev/sda1                               128M   38M   84M  31% /boot
/dev/mapper/core_vg-core                 25G  173M   24G   1% /storage/core
/dev/mapper/log_vg-log                  9.9G  1.4G  8.1G  15% /storage/log
/dev/mapper/db_vg-db                    9.9G  204M  9.2G   3% /storage/db
/dev/mapper/dblog_vg-dblog              5.0G  171M  4.5G   4% /storage/dblog
/dev/mapper/seat_vg-seat                9.9G  271M  9.1G   3% /storage/seat
/dev/mapper/netdump_vg-netdump         1001M   18M  932M   2% /storage/netdump
/dev/mapper/autodeploy_vg-autodeploy    9.9G  151M  9.2G   2% /storage/autodeploy
/dev/mapper/invsvc_vg-invsvc            5.0G  155M  4.6G   4% /storage/invsvc
```

Here, additional String manipulation can be run on the output.

Introduction to vRealize Orchestrator

vRealize Orchestrator (**vRO**) is a tool that is included with vCenter Server. The appliance can create custom workflows, talk to different systems, provide a simple workflow graphic tool, connect to, and utilize REST functions, and export configurations and controls to disparate systems, which single programming languages can't always do. Even though vRO can be a little more complicated to implement and operate, it can provide a fantastic mechanism for a DevOps system and provide a simple and central tool for all Automation Engineers.

Summary

This chapter dealt with combining PowerCLI and PowerShell, or PowerCLI and BASH, to control multiple different systems. It helped pass the control of such systems through multiple machines, through the VM structure and through remote PoSH with `PSSession`.

The next chapter shows the development of workflows, the installation of vRealize Orchestrator, and the installation of a designer, which is used to show the workflows with a graphical flair.

5
Workflows and vRealize Orchestrator

Mr. Mitchell was true to his word. The boss was given his papers, and you are heading up the DevOps initiative with a budget to get a good start. One problem, where do you start? What needs to be done first?

After a day of reflection, reading, and learning, there were very few articles on the subject. What you found was limited and not as helpful as expected. Then, the "Aha" moment hit you; what do we have already? Frank is always talking about this Orchestrator to do his VMware administration, you think. "Hey Frank, how's it going?" He responds with "Good! I am pretty busy today; what can I help you with?"

You begin by explaining your dilemma and ask where he started with his and what he does with what he develops. He lights up with pride as he explains how he got started and why he decided to script, "I ran into numerous times when clicking through things that I had too many things to do and not enough time to do it, so I started looking into it." You add, "Me too". He continues, "I am not as good as I would like to be, but I certainly see why VMware has been talking about it more and more."

This chapter is dedicated to workflows, planning them, drawing them out, and making them work. As a good visualization tool and a workflow engine, **vRealize Orchestrator** connects to many different systems allowing a centralized approach to management. It may not be a full DevOps engine but it is a start.

We will be covering the following topics:

- Workflow design
- Implementing vRealize Orchestrator
- Using vRealize Orchestrator to visualize workflows

Workflows

As in previous chapters, workflows are a process or procedure of multiple "things" in a specific order. A common misconception is that workflows are not just a script; they can and should be drawn out using a pen and paper, process flow software, or a whiteboard. It is something software developers have been doing for years and has been touched on throughout all the scripts shown in the previous chapters.

The visualization of a workflow starts at a very high level and gets deeper and deeper when more areas are fleshed out. This starts with what the user sees, leads through the expectations and outcomes, and finally concludes with the finished product; this is what a workflow will entail.

The beginning of a workflow

Talk to any process manager, software developer, or even a project manager and they understand this with their methodologies and their normal day-to-day jobs. Project managers have their Gantt charts that show tasks and time frames. Developers have their own tools that expand and explore user expectations and process managers have swim lanes and process charts. All charts and tools help with the development of a finished product or expanding an idea or need.

In DevOps or system orchestration, where would someone start? A personal perspective from numerous operational transformation engagements is to start small. Look at what is already in the script repository and start there. If the scripts are plagiarized from the Internet, so be it, expand on that. The reason these tools and scripts were downloaded or planned is because there was a need, thus the reason to expand on what is already there.

The planning stage

Basing the start on what needs already exist builds the expertise and the pool quickly. Once some of the low-hanging fruit is in place, the next job is to determine a use, a need, or identify whether it's an annoyance. A typical example is to think through a life cycle of a VM—request, provision, use, retire, and delete. This process seems simple enough until you start manually breaking out the steps for even the first item. How does a user request the VM? What process exists for consistency and speed? Imagine you have to do 200 of them in a week, is that even possible?

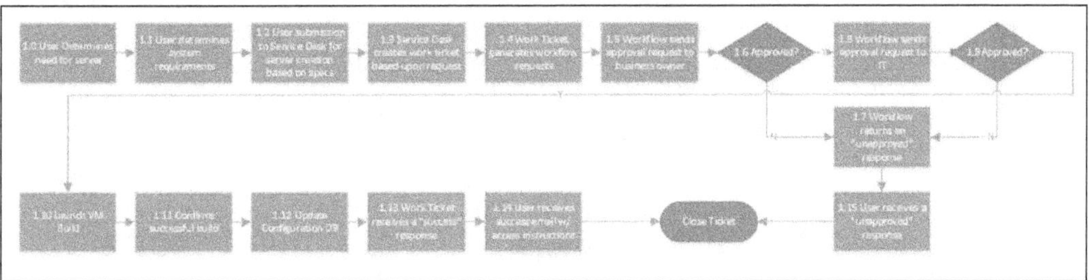

Starting and visualizing the plan creates an easier move to the next stage, which is the design stage where the individual processes will be expanding each item as shown previously. Expand the process into numbered processes to have a reference to each step, and for each step, further processes can be built upon the main process. For example, if step 1.8 needs further clarification on how to send the workflow approval to IT, such as specific users or roles in the organization, it can be expanded with another process chart with the numbering being 1.8.1, 1.8.2, and so on. Each expansion of a process item will help more and more as one moves through the phases.

The design phase

After building the processes, next comes the design. The design consists of flows and drawings similar to the preceding diagram. Using **Microsoft Visio** or **LucidChart** on a Mac will allow a visual workflow drawing to help work out the design. What will the design look like, and why draw it out? Design begins with three specific types: **Physical**, **Logical**, and **Conceptual**. Always begin with the conceptual as this really has no barriers and can be almost anything. How the communication flows, desired end state for a process, systems in use for something to occur, or a basic workflow are examples of a conceptual design. The following example is a simple conceptual design for a provisioning process where it starts from a desired state and works through to a desired end state at a particular point in time.

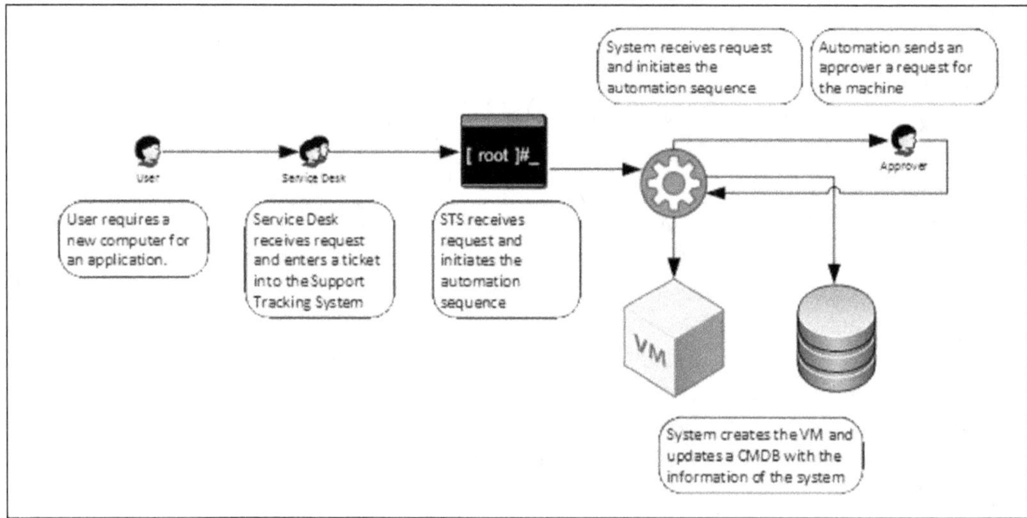

This design doesn't go into extensive detail and doesn't show how to do all of the steps but defines that an end user requires an application and a new computer to run such an application. It doesn't show a vetting process from the service desk to determine if a new computer is needed or not, what the application is or if the **Support Tracking System** (**STS**) would initiate the process from what was inputted. However, it does show the desire for automation, approval, and a **Configuration Management Database** (**CMDB**) that would store the information of the build. This is a great starting point and would be an initial draft or first cut at the conceptual design.

The second area is the **logical design**; this takes the conceptual design and fleshes out each point and puts technologies or systems in place to perform the desired state. For example, if the company is running **Remedy** for their STS, then build out the Remedy requirements to push a request to the automation system. Then say, **VMware vRealize Orchestrator** (**vRO**) was chosen for the automation engine, highlight that in the logical design and show the components capable of performing the task.

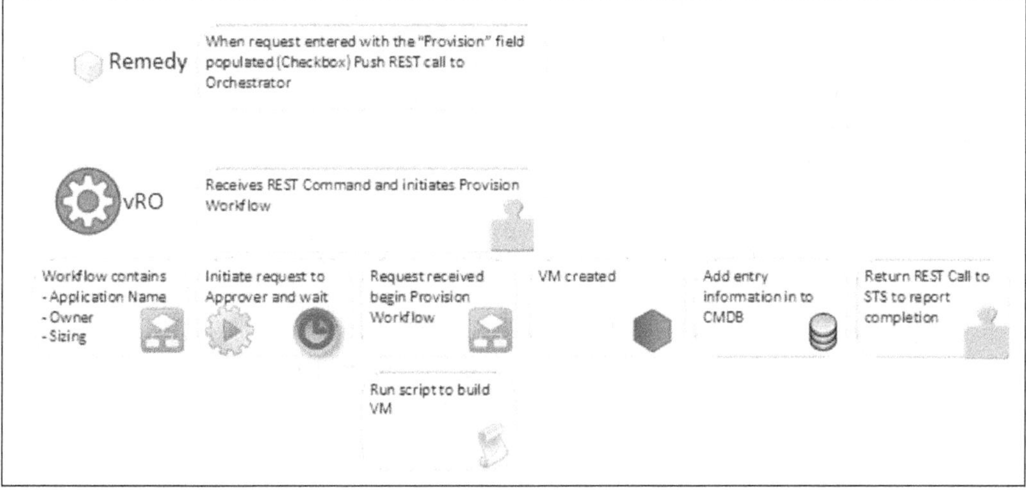

The preceding figure is a little more detailed and covers what is expected to be accomplished from **vRO**. It isn't fully baked and each piece isn't describing the complete script but puts all the components into perspective.

Physical design can be an actual physical server network port 1 to physical switch port 23 using VLAN 244, or it can be the actual writing of a script's components. It doesn't need to include pictures and is better suited to a table format or point form, and in some cases the physical design can be the script itself.

The implementation phase

The implementation phase, sometimes referred to as the deployment phase is the writing of the script, the building of the workflow in the tool, or the installation of the base system to control the automation. This is where the build occurs and depending on the time spent on the design, determines the length of time that this phase requires. Sometimes, this can be quick and quite often intermixes with the next phase of testing.

The testing phase

In the testing phase, or validation phase, the build or implementation is tested for validity, errors and/or missing functions. I have typically seen the testing phase and the implementation phase run in parallel or lockstep as the individual components are installed, tested, and returned for remediation. A testing methodology of detailed use cases is also typically employed to ensure there is a rhyme and reason to the testing.

Scenario Testing should be set up before or during the initial stages of the implementation phase. Each use case should include a desire, and expected result. For example, building a scenario based upon the figures shown in the design section; an expected result would be to have a test request in Remedy, and have a VM built upon completion. So, building variants to that use case where the name contains uppercase characters, special characters or punctuation, specifies letters in the number of CPUs, or setting the data disk outside the disk size boundaries, would be valid variants for the use case.

Promotion or closeout

The promotion or closeout portion of the workflow promotes it to production use. This usually requires documentation or how-to-use aspects of the workflow, walk-through for key users/stakeholders on the workflow and official sign-off of the completed product. If version control is being used, this is the promotion of the workflow from version 0.x to a 1.0 version of the product.

This phase is critical to any development aspect as it reports completion, a means to change the control from the development team to the operations team. This initial step needs to be smooth or the workflow, upon initial deployment, will be evaluated based on its performance.

vRealize Orchestrator

vRealize Orchestrator is an automation and workflow engine that performs remote operations on dissimilar systems for the purpose of removing human error from a task. There are other workflow engines that exist in different software platforms and perform similar types of system integration, but vRO is the best one for VMware and its various stacks. vRO began, through an acquisition from a Swiss company named Dunes in 2007, from a product named **Virtual Service Orchestrator** or **VS-O**, and became one of the best tools that no one knew about. The software would allow the graphic mapping of a process and the placement and running of scripts within the workflow. For years, there was simply no documentation or blogs about the product and it became a niche skill for the savviest Administrators.

With the purchase of Dynamic Ops in 2012, more and more Administrators were picking up the reigns of this orchestration software and Orchestrator eventually became intertwined with vCenter Automation Center (now vRealize Automation) to become the key focal point of all the separate systems' integration. Many plugins have been written for vRO and it is truly becoming a staple for organizations that need interoperability of their systems.

Why bring this tool up in a book about PowerCLI? This product has access to a number of APIs that are closed or not available within PowerCLI. It enables much more extensibility than the native PowerCLI can provide. So, to truly administer a VMware ecosystem, Orchestrator must be discussed.

This chapter will only discuss the 6.0.2 version of vRO and integrate that into the automation with the PowerCLI discussion. At the time of writing, the 7.0 version of vRO is available but is very new and certified for vRA 7.0 installation. On `www.vmware.com`, in the **Downloads** section of vSphere, 6.0.2 is the latest version of the Orchestration appliance to be shown with vSphere 6; therefore, this is the version that will be used here.

Architecture and history

For years, the Orchestrator installation was an install on a Windows machine and always on the same VM as the vCenter install itself. The following diagram is the architecture drawing from the original vCenter Orchestrator install and config guide (`http://www.vmware.com/pdf/vco_40_install_config_guide.pdf`).

It followed a fairly simple layout and design where an install of a fat client application could be run from an Administrator's workstation and was a means to automate the control using a separate interface.

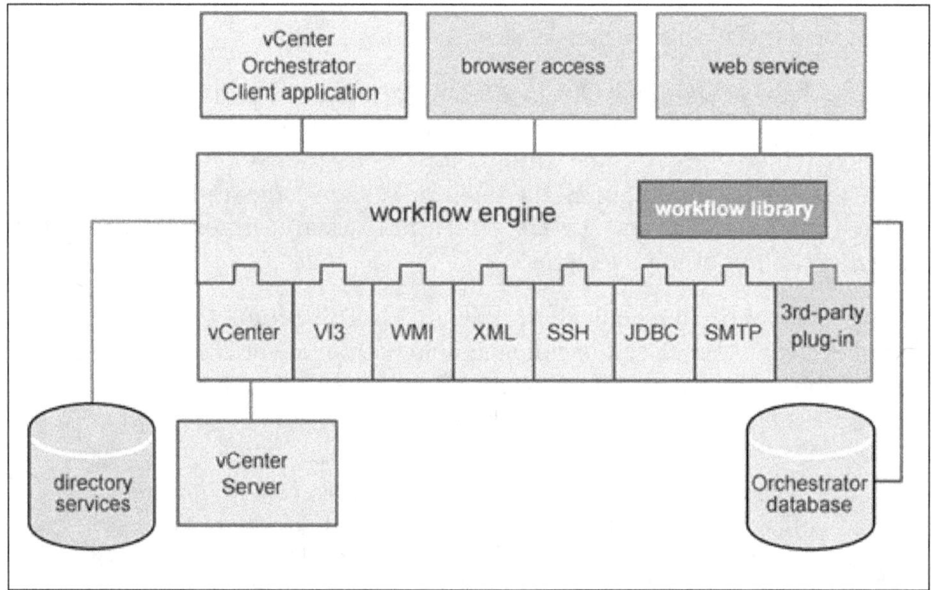

In vCenter version 5.1, there was a separate appliance that was introduced to help when the installation of vCenter was on the virtual appliance. The architecture didn't change much from the previous versions with the exception of using SSO and REST/SOAP calls. It is used from the install and configure guide (`https://pubs.vmware.com/vsphere-51/topic/com.vmware.ICbase/PDF/vcenter-orchestrator-51-install-config-guide.pdf`).

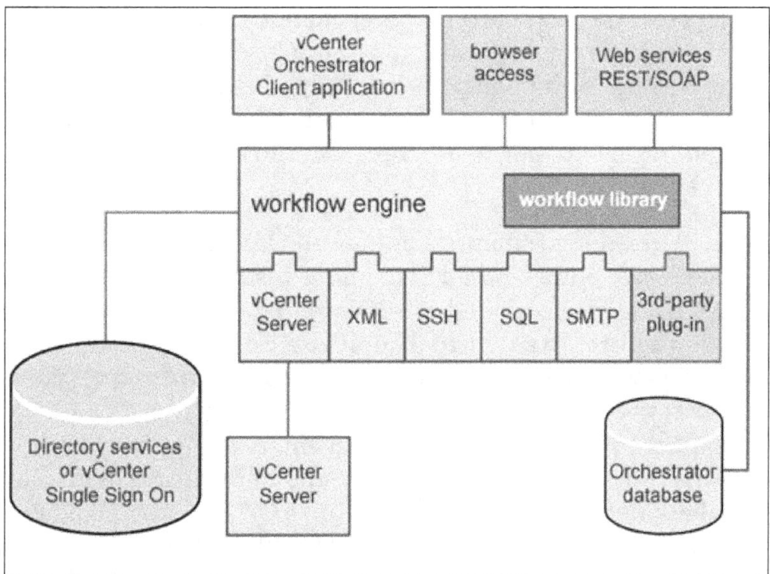

This design hasn't changed even in the latest versions of vRealize Orchestrator, with the exception that a Microsoft SQL Server database is not a requirement anymore.

The architecture for client access has also changed in later versions of the application as well. Older versions required the Orchestrator Client to perform workflows, whereas the newer versions have web portals and plugin integration into the Web Client of vCenter. So how does this fit into a typical environment?

Sample environment

Let's start with a small single site implementation, which consists of a single MS Windows-based vCenter VM, eight compute nodes, and a single DRS cluster. This sample environment will be for a small manufacturing company that is still moving into an automated, centralized model. The IT team is small, consisting of five Administrators that have a varying amount of skill. A couple of the support personnel are hands-on and provide telephone support for the office and field users—one is a web developer for their intranet and customer portal and the last two are Windows, Linux, and VMware Administrators where one is junior and one is senior. Out of the team, no one has time to implement an automation framework, let alone learn something new. Does this sound familiar?

As a consultant presenting solutions to this customer, certain things stand out. The implementation of Orchestrator has to be simple, quick, and painless and the operation has to be simple and nonintrusive; it shouldn't consume a large footprint of resources and will need to have a self-explanatory interface. Considering these requirements and the fact that there are three ways to implement Orchestrator, certain methods stand out.

With vSphere 6, there are two solutions to how an Orchestrator instance can be installed—one is a Windows-based Standalone installation that resides on a Windows server (using an embedded MS SQL server, or an Oracle instance) and another is utilizing a SUSE-based virtual appliance. Some would think that installing Orchestrator on the vCenter would be ideal using the Standalone version and this is one example that will be shown in the next section; however, I would generally only do this for a test or development environment. Although, the premise of this example is that this will be for production, and performance will be an issue. Therefore, the alternate options of using a vRealize Orchestrator Standalone installation on a separate Windows server or the Orchestrator virtual appliance would be the better choice.

Special considerations

A database is required for vRealize Orchestrator and, by default, the appliance and the standalone version are preconfigured to work with a local installation of PostgresDB. According to the **Setting up the Orchestrator Database** section on the install and configure vRO document, this generally is appropriate for small or medium organizations, although this installation also restricts the usage in a cluster mode. SQL Server and Oracle databases allow for extended functionality.

Personally, I typically recommend the virtual appliance because of licensing considerations and the simplicity of the appliance. Virtual appliances generally are quick to setup; they do not use a large footprint of resources and are quick to download and get working. Considering that this is a smaller environment and simplicity is the key, the Orchestrator virtual appliance installation with the embedded PostgresDB installation has been chosen.

Where to get it?

To get the Orchestrator appliance, connect to this URL to download the OVA from the VMware website. Here is the link and the compatibility list for 6.0.3:

`https://my.vmware.com/web/vmware/details?downloadGroup=VROVA_603&productId=489&rPId=8536`

vRealize Orchestrator	6.0.3	6.0.2
VMware vCenter Server 6.0 U1	✓	✓
VMware vCenter Server 6.0	✓	✓
VMware vCenter Server 5.5 U3	✓	
VMware vCenter Server 5.5 U2	✓	✓
VMware vCenter Server 5.5 U1	✓	✓
VMware vCenter Server 5.5	✓	✓
VMware vCenter Server 5.1 U3	✓	✓
VMware vCenter Server 5.1 U2	✓	✓
VMware vCenter Server 5.1 U1	✓	✓
VMware vCenter Server 5.1	✓	✓

Notice that the newer version still works with the older version of vCenter, and it is ideal to attempt to use the later version, just like the PowerCLI version as mentioned in earlier chapters.

Lastly, the following URL shows the **Software Developer Kit** (**SDK**) for Orchestrator, which lists key commands for developing a plugin of your very own! The key point here is that it is a reference for commands (`https://developercenter.vmware.com/web/sdk/60/vrealize-orchestrator`).

Prerequisites to installation

As with any installation of a server, planning the design and setup is critical to the final implementation. Make sure to test the server web interface and the browser versions to ensure compatibility.

Version Considerations

While I was setting this up in my lab I noticed an issue with 6.0.2 and Chrome 48.0.2564.103. The 6.0.2 version of the appliance uses an insecure Diffie-Hellman SSL public key and Chrome will not open the window for configuration. There are methods to bypass this using Firefox 43.0.4 on a Mac, whereas the Firefox on a Windows had the same issue as Chrome. Upgrading to the 6.0.3 version of the appliance fixes the issue by upping the SSL key strength to 256 bit.

On the virtual appliance version of Orchestrator, there is a configuration port for the **VAMI** (**Virtual Appliance Management Interface**) using the TCP port 5480, which allows configuration of the database, network, updating, and security components of the VM. Once the OVA is installed, this port is available for configuration.

Using the standalone installation for the example consists of setting up a Windows installation VM and ensuring a database is installed on the Guest. It doesn't have the TCP port 5480 available and requires configuration through other methods.

Installation

Installing the appliance is fairly straightforward and simple, but configuring takes a little more time. Starting with the virtual appliance installation, once the OVA file is downloaded, and vCenter is installed and running (Windows base or virtual appliance), this VM is ready to be deployed.

Using the VIClient, log in to the vCenter that houses enough resource capacity to hold the appliance. By default, the appliance uses two vCPUs and 3 GB of RAM (version 6.0.3), 12 GB of provisioned storage, and 3.75 GB is used when thinly provisioned. This is the default sizing for the appliance and isn't resizable in the configuration.

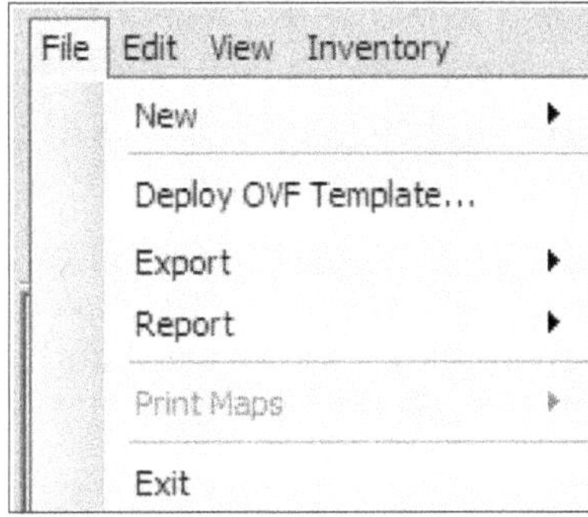

Based on the preceding diagram, let's start with the deployment process by performing the following steps:

1. Click on **File**, and select **Deploy OVF Template...** to launch the deployment process.
2. The next screen opens to show the source location for the OVA file.
3. Use the **Browse** button to locate the downloaded file, and click on **Open** to look inside the virtual appliance. Ensure that the version number and product information is correct and proceed to the next screen.
4. Make sure to read and then accept the EULA.

Product:	VMware vRealize Orchestrator Appliance
Version:	6.0.3.0
Vendor:	VMware Inc.
Publisher:	VMware, Inc.
Download size:	876.1 MB
Size on disk:	2.2 GB (thin provisioned) 12.0 GB (thick provisioned)
Description:	Automate tasks for VMware vSphere and enable orchestration between multiple solutions. VMware vRealize Orchestrator allows administrators to capture their best practices and turn them into automated workflows.

5. The next screen shows the descriptive name that the appliance is going to be called. Consider shortening the name for your benefit; we all know that "VMware vRealize Orchestrator Appliance" is a very descriptive and well-understood name, but it is really long.
6. Select the **Inventory** location, **vCenter**, **Datacenter**, and the **VM** folder to place the appliance in and click on **Next**. Over the next four configuration screens, ask for the position where the VM will reside and in which DRS Cluster, then select the storage location, the disk format, and the network connection.

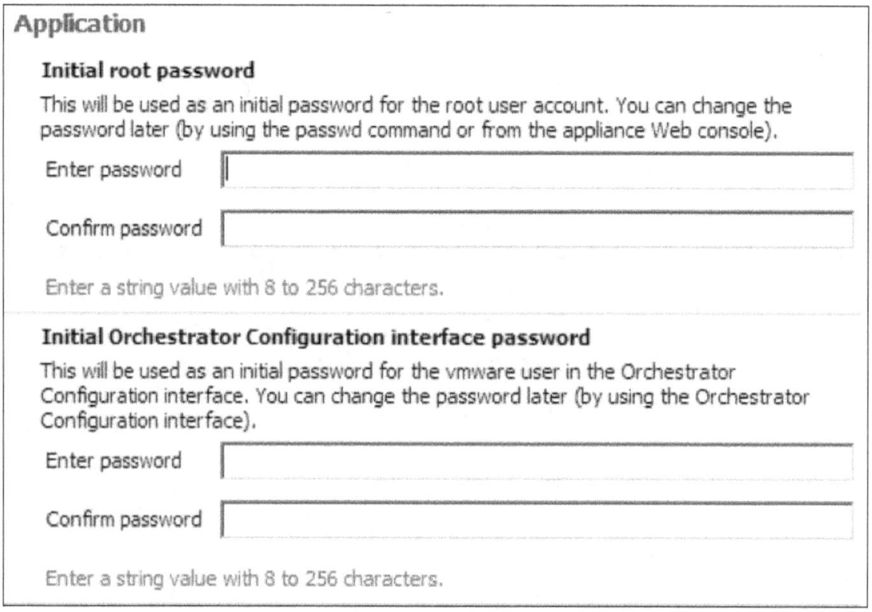

Application

Initial root password

This will be used as an initial password for the root user account. You can change the password later (by using the passwd command or from the appliance Web console).

Enter password

Confirm password

Enter a string value with 8 to 256 characters.

Initial Orchestrator Configuration interface password

This will be used as an initial password for the vmware user in the Orchestrator Configuration interface. You can change the password later (by using the Orchestrator Configuration interface).

Enter password

Confirm password

Enter a string value with 8 to 256 characters.

7. Lastly, set the root password and the Orchestrator interface password, and remember these because they will be used right away in the configuration.

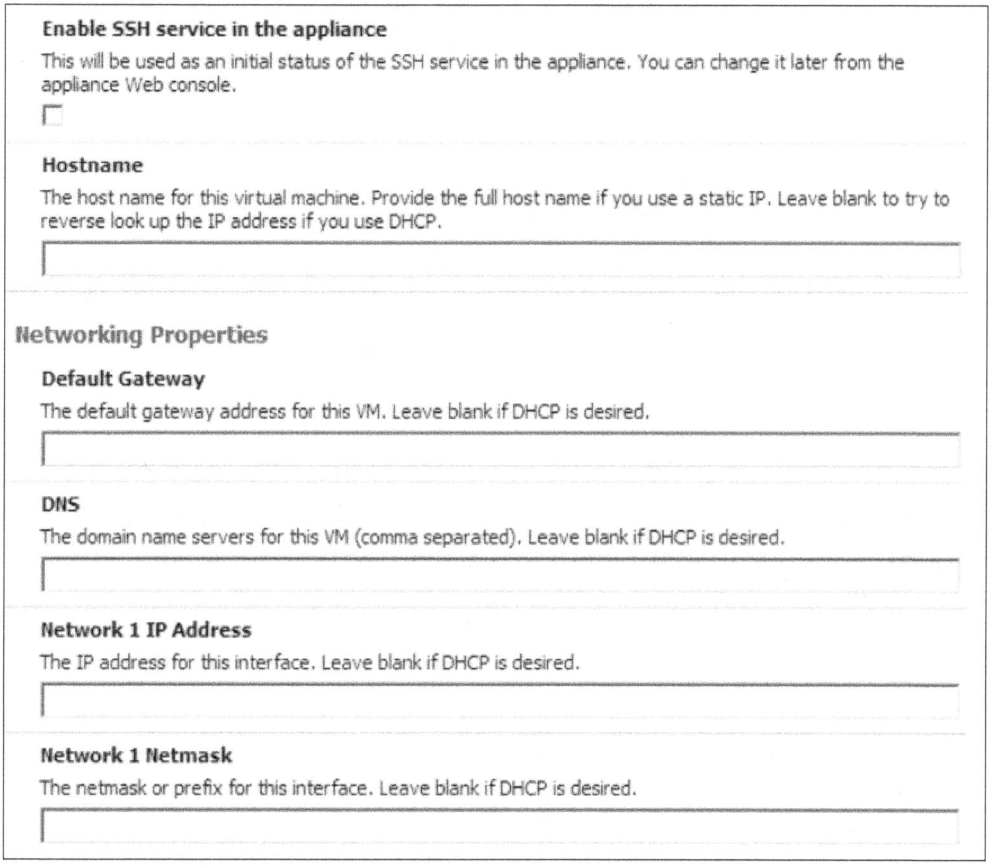

Enable SSH service in the appliance

This will be used as an initial status of the SSH service in the appliance. You can change it later from the appliance Web console.

Hostname

The host name for this virtual machine. Provide the full host name if you use a static IP. Leave blank to try to reverse look up the IP address if you use DHCP.

Networking Properties

Default Gateway

The default gateway address for this VM. Leave blank if DHCP is desired.

DNS

The domain name servers for this VM (comma separated). Leave blank if DHCP is desired.

Network 1 IP Address

The IP address for this interface. Leave blank if DHCP is desired.

Network 1 Netmask

The netmask or prefix for this interface. Leave blank if DHCP is desired.

Make sure to scroll down on this screen to enable SSH and set the hostname to be used. Also, set **Networking Properties** by entering the **Default Gateway**, **DNS**, **Network IPv4 Address**, and **Netmask** for the appliance.

Upon completion of this last configuration screen, as in most OVF deployments, there is a display for the configuration items chosen during the previous wizard. This allows the confirmation of the settings before deployment. Click on **Finish** to begin the installation.

Utilizing the Web Client has a different look and feel around the install, as one logs in at `https://<vcenterServer>:9443/vsphere-client`.

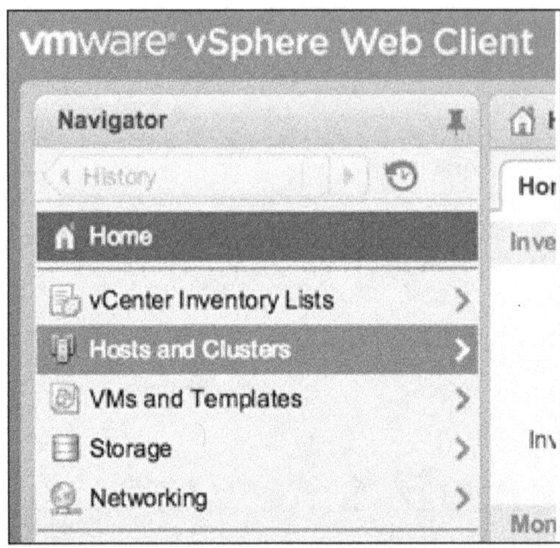

8. Select the **Hosts and Clusters** option on the left pane. Select the vCenter to place the appliance in and right-click on it, exposing a menu in which **Deploy OVF Template...** can be selected.

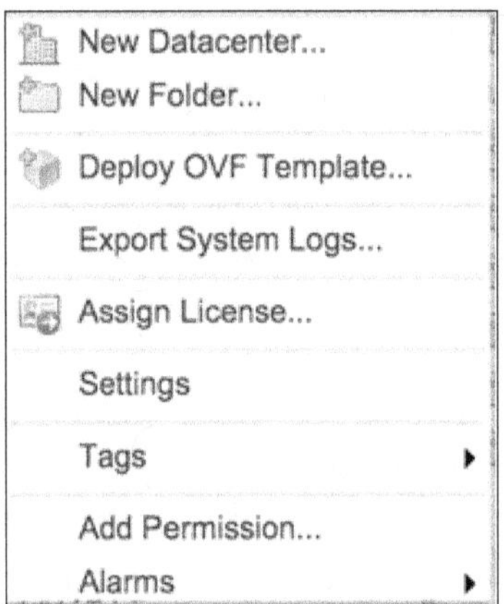

The Web Client gives the opportunity to use a URL or a Local file as the VIClient. The URL can get the OVF package directly from the Internet, if you know the path to the file.

The familiar **Review details** of the OVF is on the next selection screen.

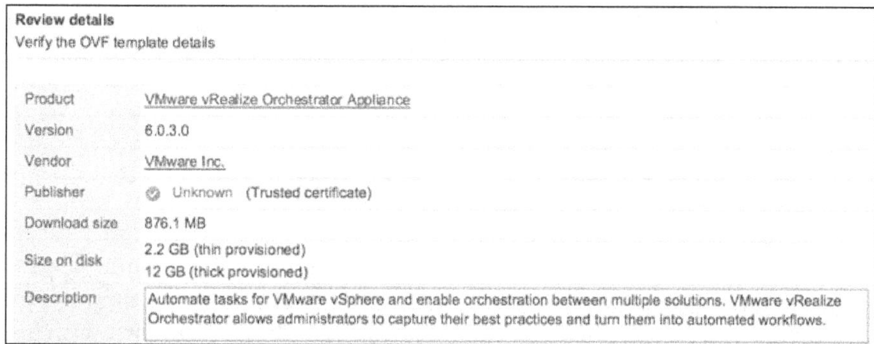

The same EULA is shown, the same VM name, and folder/datacenter selection with the difference here being a **Search** field to find a location for a large environment. Select the resource (**DRS Cluster**) with the same **Search** field, and select the storage that is much better laid out with the disk format and the location shown on one screen. The Datastore locations also show much more information for the Administrator to select a better location for the appliance.

The network setup validates the configuration and allows an IPv6 configuration to be used. The customized template screen now shows error icons if invalid entries exist.

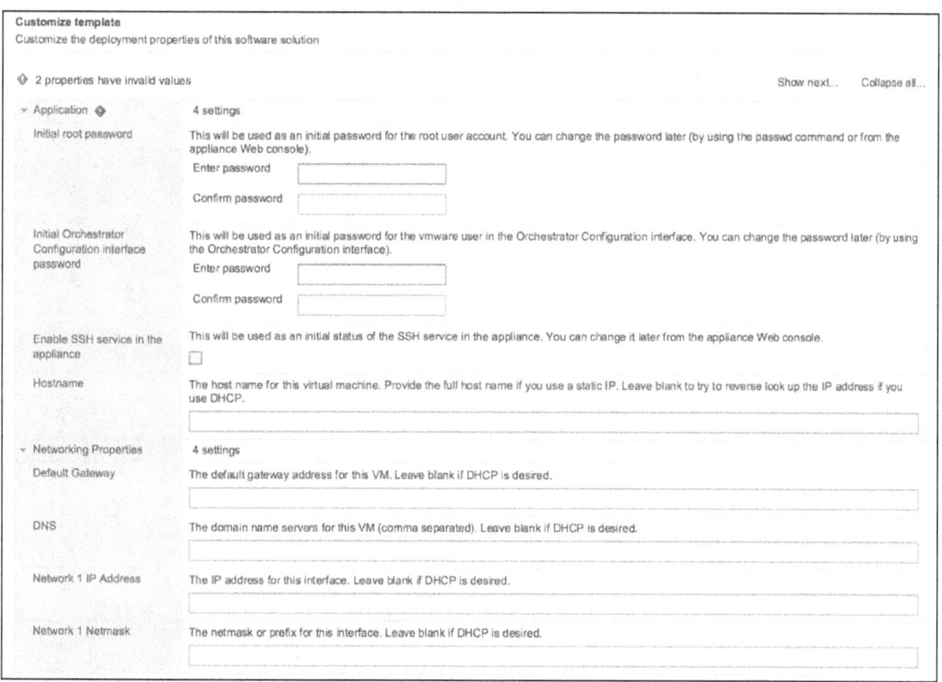

However, the information required is still the same. This run-through also stops the **Next** button from being clicked on without a proper password being entered into the field, whereas the VIClient one would allow the **Finish** button to be clicked on with a blank entry. Once confirmed and run, vRealize Orchestrator Appliance will be installed.

Installing the standalone instance

Using the current version, 6.0.3, the Windows base standalone installation is available, whereas the version 7.0.0 standalone instance is not available. There is good reason for that from the VMware front as virtual appliances generally are simpler to install and are simpler for VMware Global Support to work with and provide a consistent platform for the products. With the development of vRealize Automation 7.0, certain backend changes occurred and simplifying the deployment helps stabilize the whole system.

Installing the standalone version is pretty easy as well. Let's look at the following steps:

1. Start with the `vRealizeOrchestrator-6.0.3.exe` file and launch it by double-clicking on it.

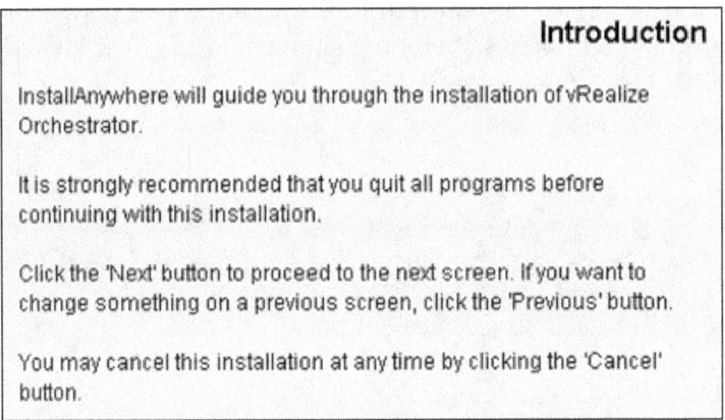

2. Choose the folder where you want to install it:

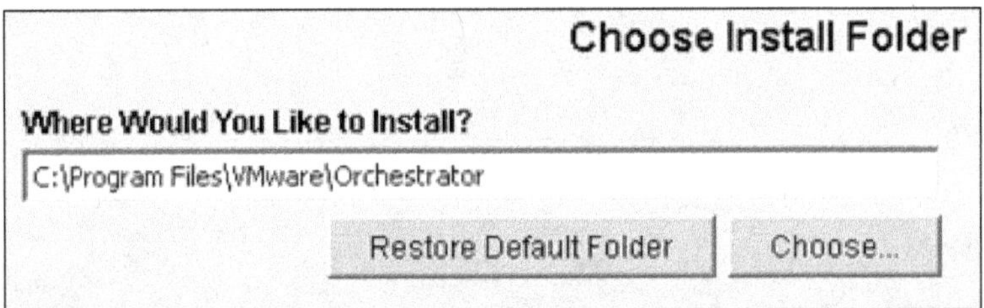

Accept the EULA and proceed to the **Install Set** selection. **Client** is the install client that allows direct interaction with Orchestrator; this will be used extensively in later sections of the book, but is not required here. The Server module is the desired install component as this is the controlling portion that is needed for Orchestrator to run.

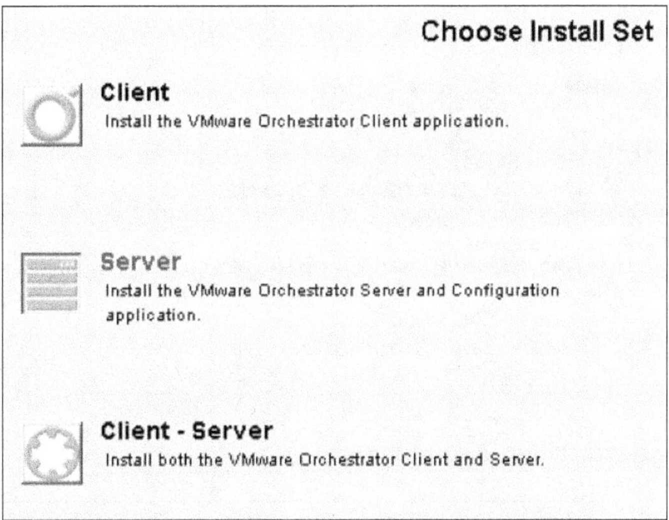

3. Set the location of the shortcut in the **All Programs** folder for itself or with other VMware products.

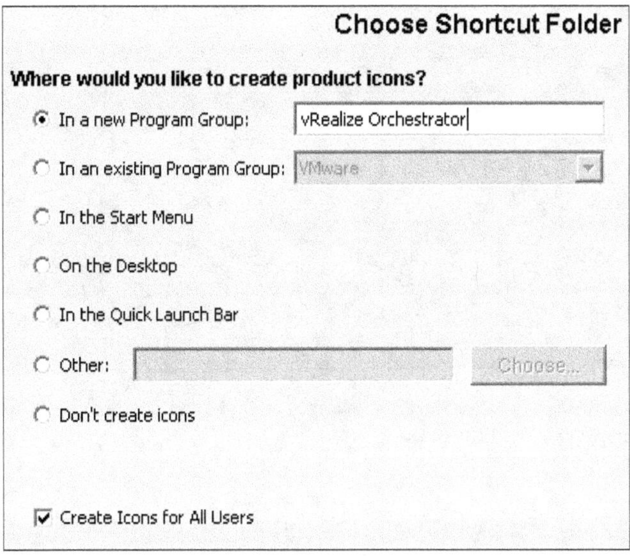

4. View the installation configuration and click on **Install**.

5. Wait for the installer to complete.

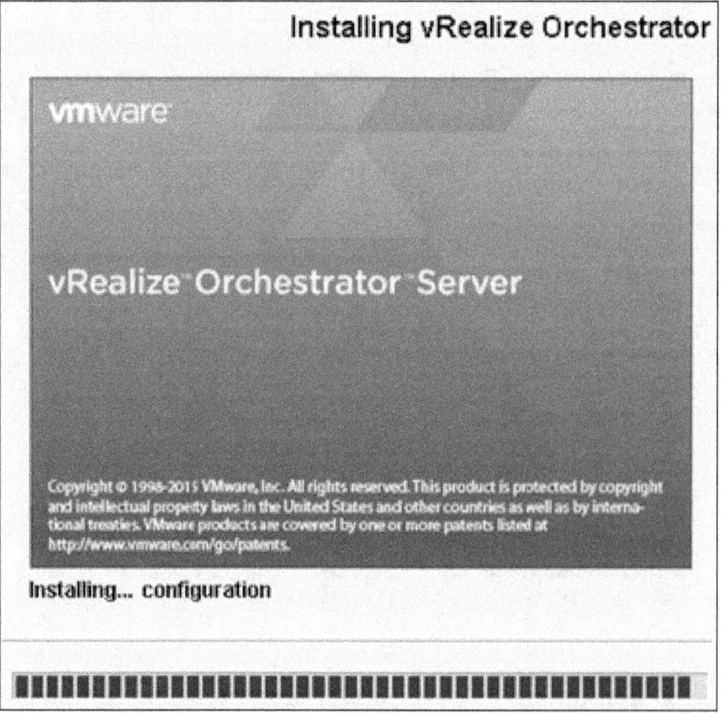

Configuring and integration

This section was the one that I personally found frustrating as I saw very little documentation that walks through every initial setup of the appliance. The configuration has changed quite a few times throughout the versions. Once the appliance is deployed, the next component to configure is the VAMI interface. Remember when I pointed out to record the password information in the install section above? Now log in with the root account and the password to edit and verify the **Time Zone** and **Network** configuration. The first screen shows the version number and the hostname and it also shows the **Reboot** and **Shutdown** actions.

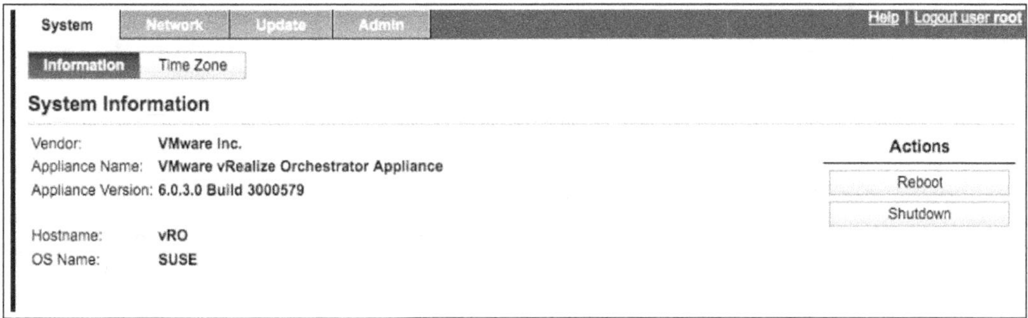

Click on **Time Zone** and set the appropriate time zone for the area and ensure you click on **Save Changes**. Under the **Network** tab is the configuration of the VM itself and some pieces can be configured under the **Address** button. The **Update** tab allows an update of the appliance, which in this case should not be used (as there is no desire to move to version 7.0.0), and lastly, the **Admin** tab allows changes to the root account password.

Once these are configured it is time to configure Orchestrator itself. In the browser window, go to the address `https://<vROServer>:8281/vco/` and there is the **Getting Started with vRealize Orchestrator** link page.

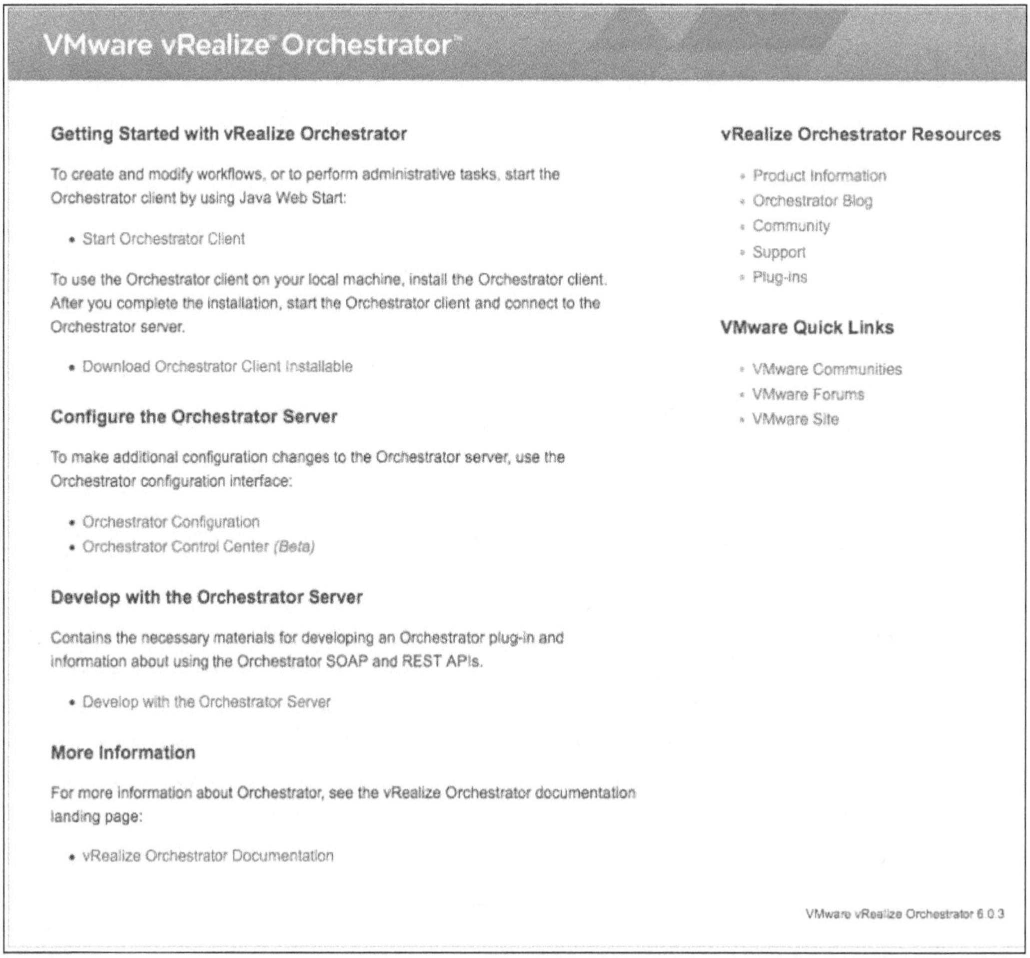

VMware vRealize® Orchestrator™

Getting Started with vRealize Orchestrator

To create and modify workflows, or to perform administrative tasks, start the Orchestrator client by using Java Web Start:

- Start Orchestrator Client

To use the Orchestrator client on your local machine, install the Orchestrator client. After you complete the installation, start the Orchestrator client and connect to the Orchestrator server.

- Download Orchestrator Client Installable

Configure the Orchestrator Server

To make additional configuration changes to the Orchestrator server, use the Orchestrator configuration interface:

- Orchestrator Configuration
- Orchestrator Control Center *(Beta)*

Develop with the Orchestrator Server

Contains the necessary materials for developing an Orchestrator plug-in and information about using the Orchestrator SOAP and REST APIs.

- Develop with the Orchestrator Server

More Information

For more information about Orchestrator, see the vRealize Orchestrator documentation landing page:

- vRealize Orchestrator Documentation

vRealize Orchestrator Resources

- Product Information
- Orchestrator Blog
- Community
- Support
- Plug-ins

VMware Quick Links

- VMware Communities
- VMware Forums
- VMware Site

VMware vRealize Orchestrator 6.0.3

Under the **Configure the Orchestrator Server** section, there is a link for **Orchestration Configuration** that sets the connection to vCenter, the database, and the client. Once the link is clicked, it takes the user to the login page for Orchestrator.

This was the frustrating part for me as the login and password defaults have changed over the years and over the versions. One version has `vco-admin`, others state the `vmware` (username) - `vmware` (password) pair, and even others have reported something else. Well, in version 6.0.3, the username is `vmware` and the password is the text that was set up during the OVF deployment.

The configuration off the bat shows all green lights, as if everything is configured and ready to go. This is a bit misleading as the vRO installation is not connected to anything and not doing anything; it requires a few key changes to this screen.

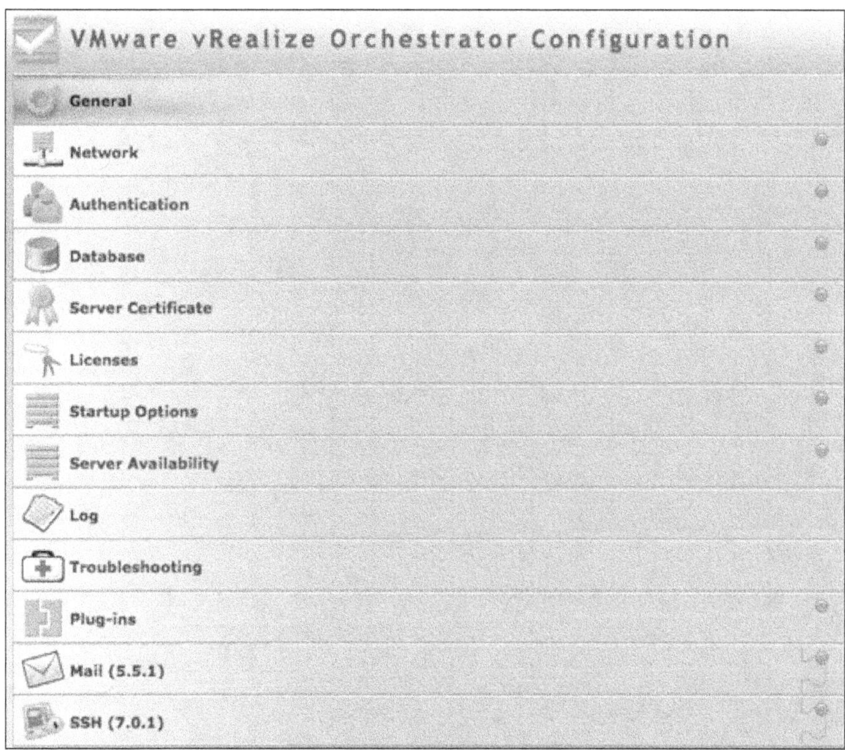

Starting with the **General** action, there is a listing of tabs on the right hand pane. These are where the key configuration needs to take place. The key point here shows a component registry and the infrastructure node information being requested, so take note that this tool is being extensively used when **vRealize Automation** (**vRA**) is in use. In the instance that this book and section are using vRO, it is a means to run PowerCLI scripts within a straight vCenter environment and not as a vRA component. Therefore, the General action will not be used at this point in time.

Next is the **Network** action, which contains two tabs, the **Network** tab and the **SSL Trust Manager** tab. Click on the **Network** tab, select the IP address to use (the DNS name should automatically change), and see the **Lookup Port**, **HTTP server port**, and the **HTTPS server port**, and keep note of that for future record. The SSL Trust Manager will be used later.

Under the **Authentication** action, there is **Authentication** and **Test Login**. The **Authentication** tab allows different authentication stores to be used: Active Directory, Local authentication, and an OpenLDAP server or VMware SSO. In the case of the example, this section will be using **VMware SSO** as it is simple but useful. By default, the authentication mode is **LDAP authentication** but is pointing to a local instance, as follows:

Authentication

Authentication mode: LDAP Authentication

LDAP client: OpenLdap

Primary LDAP host: localhost

Secondary LDAP host:

Port: 389

Root: dc=appliance

Use SSL

URL ldap://localhost:389/dc=appliance

User name: cn=vcoadmin,ou=vco,dc=appliance

Password: ********

User lookup base: ou=vco,dc=appliance Search

Group lookup base: ou=vco,dc=appliance Search

vRO Admin group: cn=vcoadmins,ou=vco,dc=appliance Search

Request timeout (ms): 5000

Dereference links

Switching **Authentication mode** to **SSO authentication** allows pointing to the SSO or **Platform Services Controller** (**PSC**) server and a common Admin user.

Authentication

Authentication mode: SSO Authentication

SSO Registration - Basic Mode

Host:

Advanced settings

Admin user name: administrator@vsphere.local

Admin password:

Register Orchestrator

SSL Certificates

Switch **Authentication mode** to **SSO Authentication**. Enter the PSC's fully qualified domain name in the **Host** field, add the administrator@vsphere.local account in **Admin user name**, and enter the password for the account. Click on the **Register Orchestrator** button to register.

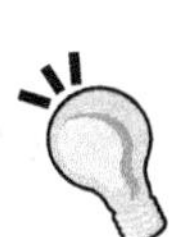

Service Enhancement

Ideally, clicking on the **Advanced Settings** link on the right-hand side allows some additional information. After clicking on **Register Orchestrator** it will be shown fairly quickly that the URL is incorrect and that the SSL certificate is not registered. There is a fairly easy fix; near the bottom of the page is a link for SSL certificates that returns the interface back to the Network Action and the SSL certificates. Make sure to copy the Lookup Service URL https://<psc_server>:7444/sts/SRSService/vsphere.local from the **Authentication** screen and click on that SSL Certificates Link. Enter the preceding URL into the **Import from URL** text box and hit **Import**. This adds the PSC Certificate to the vRO configuration.

The next part of setting the SSO Domain is important to the tool itself. This portion will set who the vRO admin user or group is. In the example's case, Active Directory is ideally where the permissions were set for vCenter instead of using local SSO accounts. Therefore, click on the **SSO Domain** dropdown and select the **Active Directory** domain in the list of domains. The **Groups filter** field allows a means to shrink the AD groups from the **vRO Admin - domain and group** selection dropdown.

Once the group or user is selected, click on **Accept Orchestrator Configuration**. The Authentication status light turns from red to green as the configuration is accepted.

The remaining actions are extra to the configuration as the Orchestrator install is functional at this point. However, it is good to cover at a high level the remainder of the actions as follows:

- **Database**: This specifies the database to be used. By default, PostgreSQL is chosen; by default username and database name use `vmware` and it cannot be placed in **Highly Available** (**HA**) or Cluster mode using this database type. For most organizations, this is an acceptable risk as this is not a critical system. If this risk is not tolerable for your company, consider the available HA modes as outlined in the VMware documentation and these knowledge base articles at `http://kb.vmware.com/kb/2079967` and `http://kb.vmware.com/kb/2118344`.
- **Licenses**: The licensing model of Orchestrator is associated with the vCenter license. In the **Licenses** action, there are a couple ways to add the license—through pointing it to a vCenter server or through a manual process.

  To enter the license through the vCenter license, add the server name, username, and password.

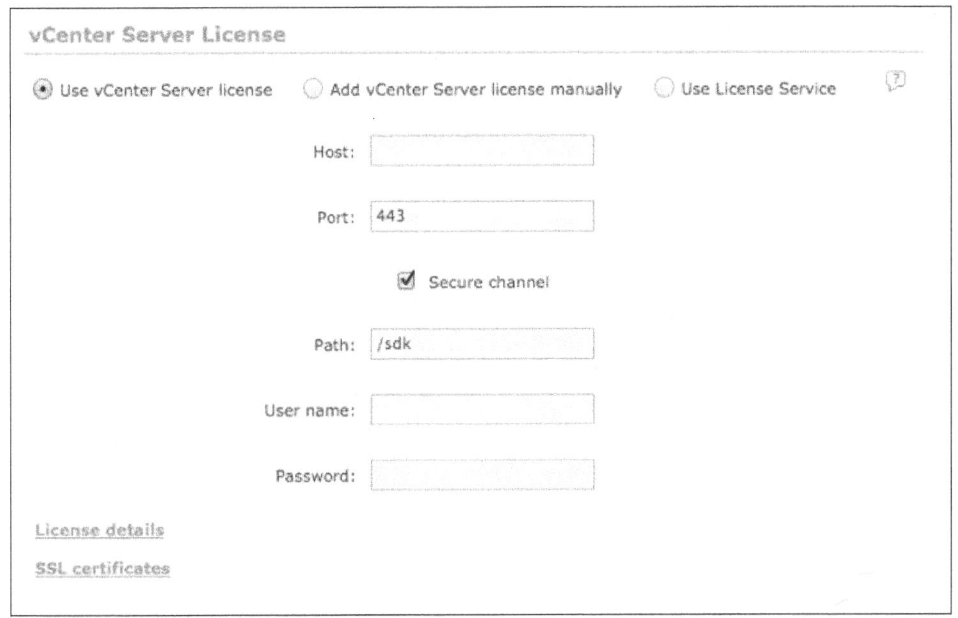

To add through the manual process, just enter the license key and the owner.

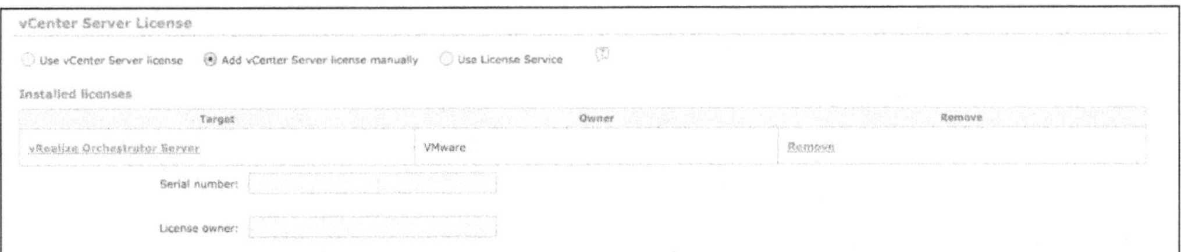

- **Startup Options**: This is essentially where the vRO server service is restarted.
- **Server Availability**: This allows the configuration of the cluster mode versus the standalone mode. See the kb articles shown previously in the database section for cluster mode considerations.
- **Log**: This is the location of the Log files.

- **Troubleshooting**: This is a great little area if the orchestrator server is "stuck" or is having problems. This action area is great for stopping the processes from overrunning the server itself.

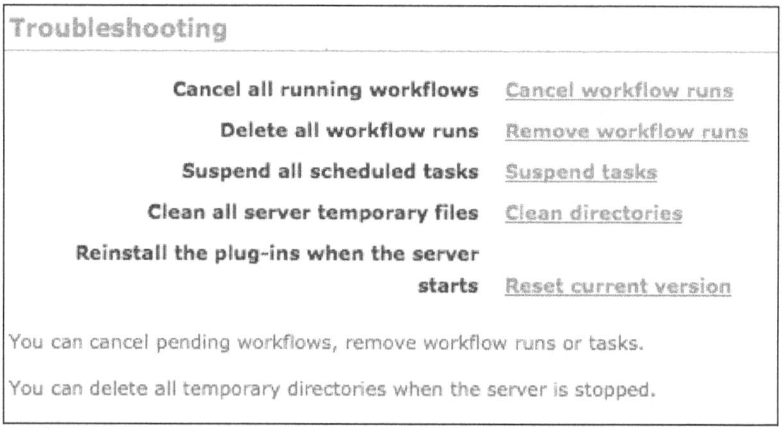

Troubleshooting

Cancel all running workflows	Cancel workflow runs
Delete all workflow runs	Remove workflow runs
Suspend all scheduled tasks	Suspend tasks
Clean all server temporary files	Clean directories
Reinstall the plug-ins when the server starts	Reset current version

You can cancel pending workflows, remove workflow runs or tasks.

You can delete all temporary directories when the server is stopped.

- **Plug-ins**: These are the installed plugins in the system. By scrolling through the installed/enabled plugins, you will see PowerShell 1.0.7.2984800 installed. This is the latest version of the plugin and will be the key component in which to run PowerCLI scripts through Orchestrator in the next chapter.
- **Mail**: This is a plugin within Orchestrator and can have the notification values defined here.
- **SSH**: This is another plugin that is configured within the orchestrator tool itself.

Once this is all complete, the next function is to enable access to Orchestrator.

Firstly, restart the vRO Server as this ensures that the new authentication changes, recently performed, have taken affect.

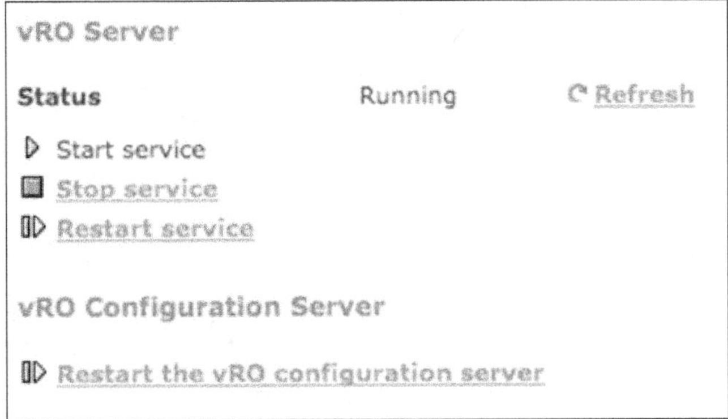

Next is the Orchestrator Client, where a web-based Java window can be opened:

Getting Started with vRealize Orchestrator

To create and modify workflows, or to perform administrative tasks, start the Orchestrator client by using Java Web Start:

- Start Orchestrator Client

Otherwise, you can download and install the installable client.

To use the Orchestrator client on your local machine, install the Orchestrator client. After you complete the installation, start the Orchestrator client and connect to the Orchestrator server.

- Download Orchestrator Client Installable
  - vRealizeOrchestratorClient-64bit-6.0.3.exe Windows 64-bit
  - vRealizeOrchestratorClient-macosx-6.0.3.zip Mac 32 and 64-bit
  - vRealizeOrchestratorClient-64bit-6.0.3.bin Linux 64-bit

I personally find the installable client works a bit better but it is a matter of the version of Java installed and running on your browser.

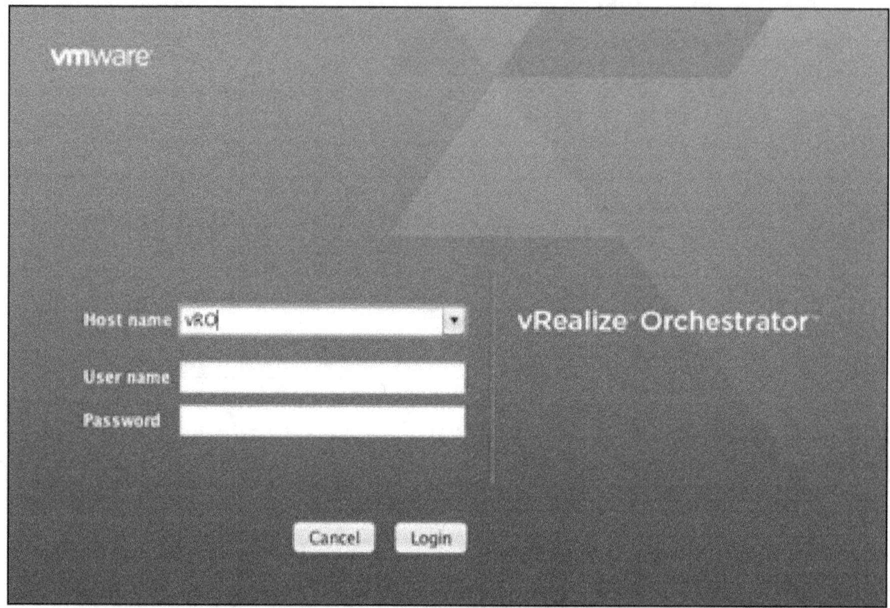

Enter the **Host name** of the vRO server, the domain-based user account, and the password, and then click on **Login**.

Finally, the vRO installation is up and running.

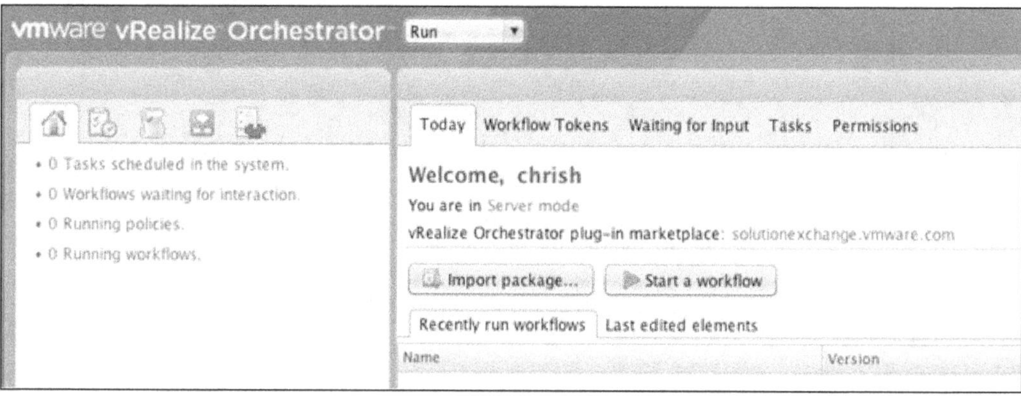

Summary

This chapter outlined process development, developing the process for the workflows, and installing and setting up **VMware vRealize Orchestrator** (**vRO**). This foundational chapter will be the key for building a strong DevOps framework and will be built upon in the next chapter.

In the next chapter, the key focal point will be to allow PowerCLI to be able to enhance the life of the Administrator, improving the connection points between the Administrator, vCenter, and the rest of VMware's virtualization stack.

6
Running Workflows with Other VMware Products

After almost 3 months, the DevOps initiative using vRealize Orchestrator is speeding things up significantly. Mr. Mitchell, up to this point, has been pleased with the progress you and your group have been able to achieve and has been touting your team to the rest of the business units. The added pressure upon your, once small, agile team is beginning to show with bogged down deployments due to other teams, extensive changes outside of your control, and unrealistic customizations demanding application integration with your processes. In building a lifecycle management, one of the key requests is to control the build and the decommissioning process and with that, take into consideration which machines are in use or not, who owns it, and who requested it.

Your team is beginning to complain stating, "The network team is taking up to a week to process our requests, and we are being blamed for it!" and "The Virtualization team is asking for scripts to manage patching and monitoring. Should we be building this stuff for them?".

Orchestrator is currently able to keep up with the demand but the rest of the product stack is woefully falling behind, as the other teams haven't adopted the "Automate First" mentality and Mr. Mitchell is asking you to step in. You are starting to look at vRealize Automation, private and public cloud offerings, this open source OpenStack product, and now on top of that, some of your key automation engineers are running out of new and interesting work and beginning to look at other companies. Automation helps but the administration, integration, and keeping track is slowing the process down and soon the frustration will leave you with a different problem than that Friday afternoon script.

Automation is definitely the way to help the working of a small and agile team; however, integration and administration will eventually move up the stack. Other teams will get in the way and eventually slow down the whole process again. This chapter deals with a multitude of VMware products and their integration with PowerCLI, Orchestrator, and keeping track of the builds and decommissioning of VMs and services. It will help develop a mindset of building at the speed of the business.

In this chapter, we will cover the following topics:

- Continuing to use vRealize Orchestrator
- Using PowerCLI to integrate NSX
- Site Recovery Manager
- vRealize Operations Manager
- VSAN

Orchestrating VMware products

Using vRealize Automation or vRealize Orchestrator is great for automating and orchestrating technology verticals or certain product integrations, but sometimes, it is just great to think low level by scripting a one-line script in certain situations to get a quick answer. As this section delves into integrating with other VMware products, I get a chance to talk about how each works, the importance of the technology stack, and some short examples of how to script in it.

Orchestrating with Orchestrator

After the last chapter where vRO was installed, the next point is to configure and run Orchestrator. We left off at the **Welcome**, **Username**, and **Run** sections of the product. In the left pane, there are five icons that perform certain actions, explained as follows:

- The **Home** tab is the starting place of the Orchestrator client. It allows the running of previously run scripts, gives a view of running policies or workflows, and scheduled task list, and gives a link for getting more plugins and packages. There is more, but it is irrelevant until there is more knowledge of the system.

- The second tab is **Scheduler**; it displays scheduled workflows that are running. At this point, there isn't anything to see in this tab.
- The third tab is **Policies**, which will be described later in this section.
- The fourth tab is interesting for a beginner to this tool. It is **Workflows**. This is the tab where the majority of users will spend more time in.
- The last tab is **Inventory**, which shows the attached plug-ins to the system and provides a version number and status for each one.

Workflows

Although this book is about PowerShell, I will stick to the PoSH-based workflows and how to set them up. By default, the plugin comes with these built-in workflows:

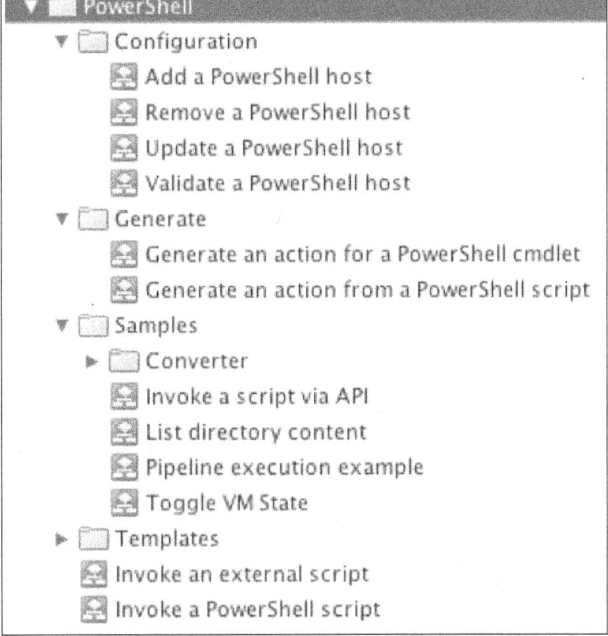

Let's start with a PowerShell Host; we'll perform the following steps:

1. Click on **Add a PowerShell host**, but what do we do with it? How do we run it? In viewing the screen, there are a few things that can be clicked on, but generally, starting with the **Run** icon invokes good feelings when working in IT.

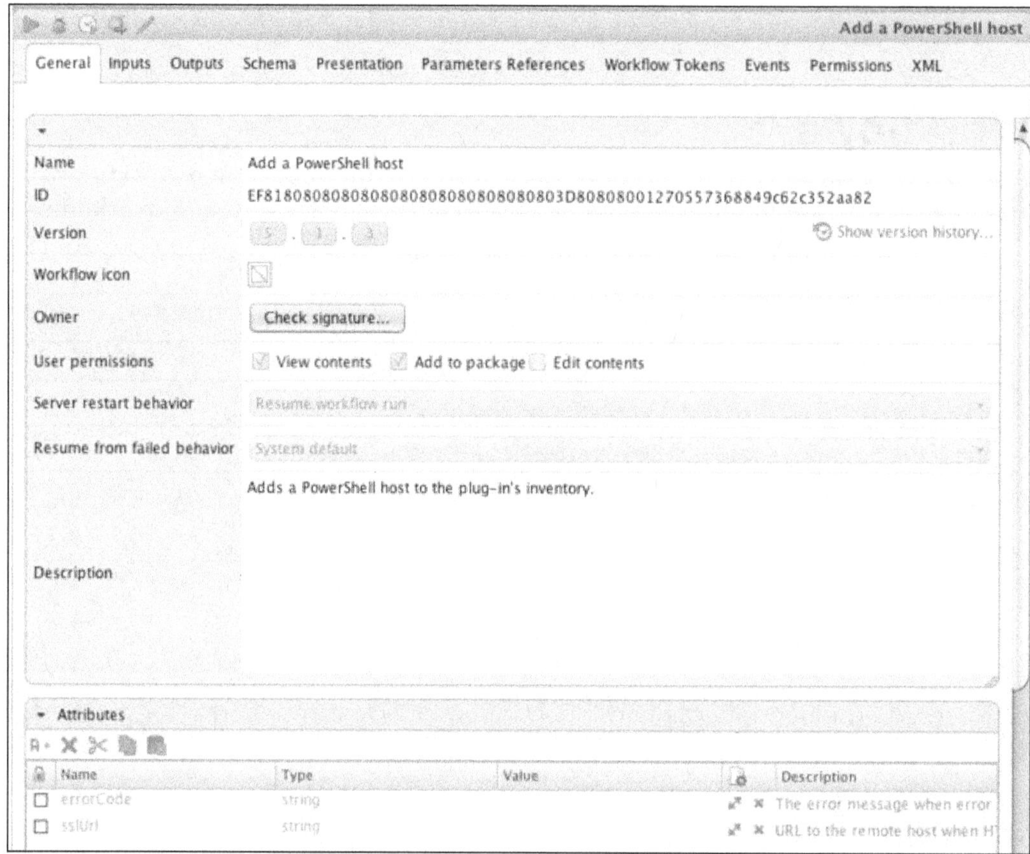

2. In the top-left corner of the screen, there is a tiny green arrow that when scrolled over displays the **Start Workflow** comment. Clicking on this will begin the workflow:

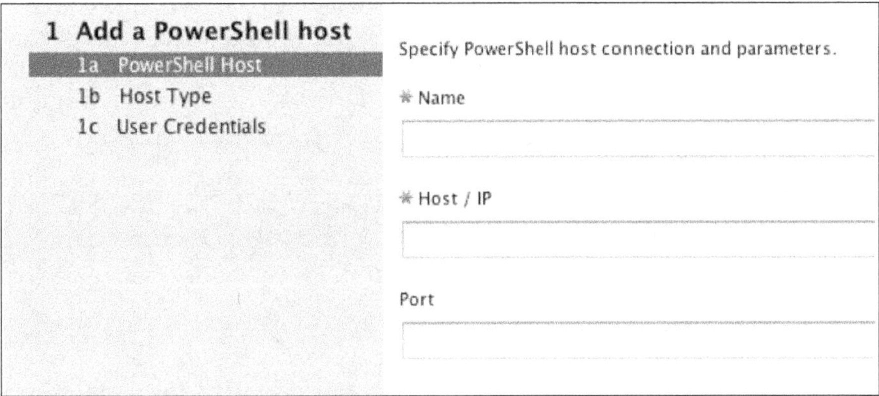

3. Enter the name of the ScriptHost, the **Host/IP** of the machine to connect to, and the **Port**, if it is not standard.

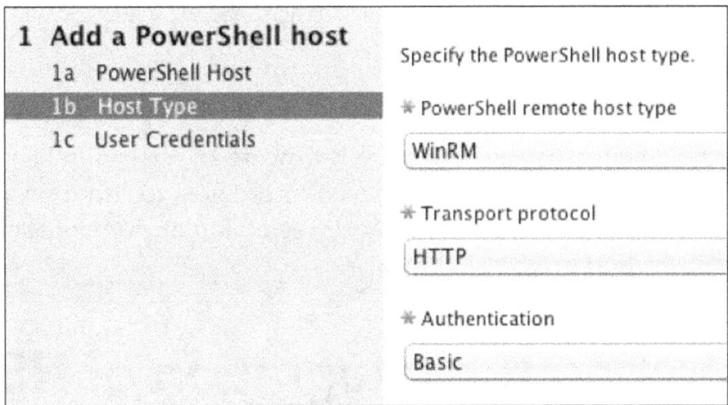

4. Choose the remote **host type**, **Transport protocol** (use **HTTP** by default), and **Authentication**.

WinRM and a PowerShell Host

When attempting this for the first time, there were a number of complications in connecting to the ScriptHost. Primarily, the WinRM `quickconfig` command didn't completely set up the Windows 2012 R2 VM that was being used. After some searching, I found a link on the VMware blog site that showed some commands that can be used on the script host.

`http://blogs.vmware.com/orchestrator/2011/12/vco-powershell-plug-in.html`

If **Shared Session** is selected, the provided credentials are used. If **Per User Session** option is used, the credentials are retrieved from the currently logged-in user.

* Session mode

Shared Session

* User name

* Password

5. Lastly, set the session mode, username, and password.

Once the script host is in place, a PowerShell/PowerCLI script can be run. Launching the **Invoke a PowerShell Script** function asks for the host to run it on and the script to run. This is a great way to populate or add a selection of premade scripts into a centralized Orchestration tool.

Orchestrating the integration of NSX

VMware NSX is a network virtualization product that allows a company to pull the configurations of each network port, firewall rule, or virtual LAN out of the realm of complexity and into an overarching framework. Network virtualization, as a concept, isn't overly new; however, what is being done with the theory is a change in thought and substance.

NSX conceptual overview

NSX was an acquisition of a company named Nicira in 2012. The Nicira product looked at an entire network and attempted to figure out a way to offer a virtualized or abstract network on top of an existing physical network. Virtualized infrastructure offered a means to the overall solution.

In a standard infrastructure, physical computers contain hard drives, CPUs, Memory, and Networking. The computer loads an operating system, such as Windows or Linux, which allows an operator to perform functions and tasks. The operating system provides a transport of the tasks and functions to interact with the physical components. The VMware infrastructure provides a shim or translation layer that interrupts the transport and tricks the operating system to think it is directly talking to the hardware, but it is actually talking to another piece of software that does the exact same thing down to the hardware.

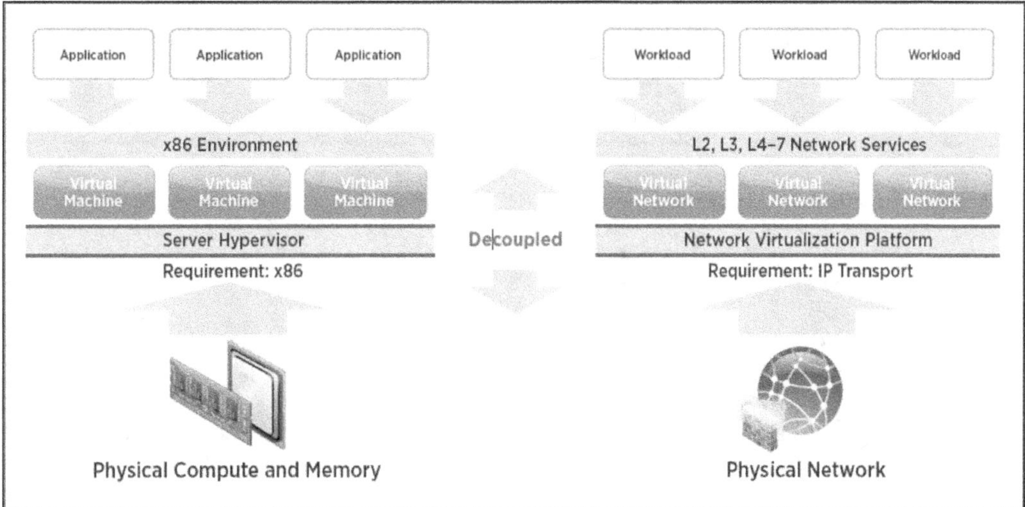

The advantage of this setup is that multiple operating systems can be installed on the same physical machine creating a means to improve the capacity by putting two or more logical or virtual machines on the same physical computer. NSX does a similar translation but with networks.

A VM is communicating to another VM on the same host but over a different subnet. Take, for example, a VM that has an address of `10.1.1.10` and is attempting to communicate to a VM with an address of `172.22.9.15`. These VMs can be on the same physical host but have to communicate up through multiple layers of switching and routing just to be returned to the same host.

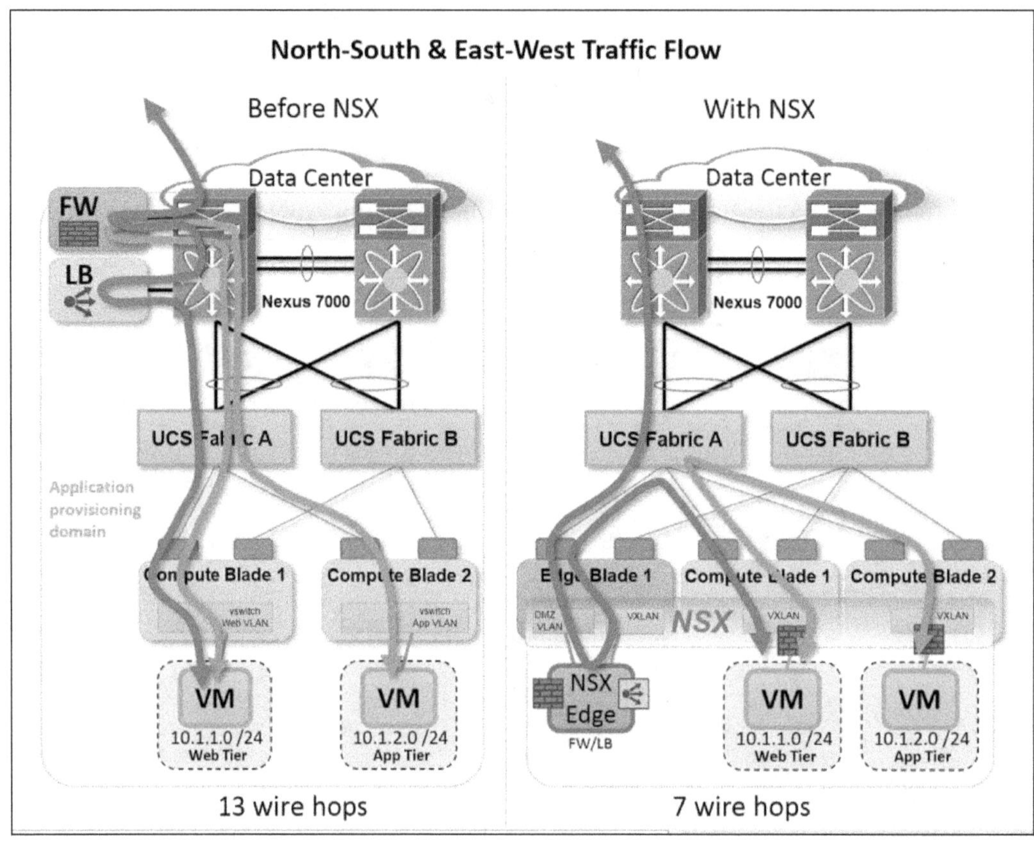

This path is one of the typical connections between the two VMs. NSX moves the communication back into the virtual switch and can speed up the processes of North-South communication and with that, allow additional services to be provided in the process. With this decentralized view of communication, the transmission of the information can be manipulated, providing solutions such as firewalling, load balancing, or application localization all within the confines of this virtual network.

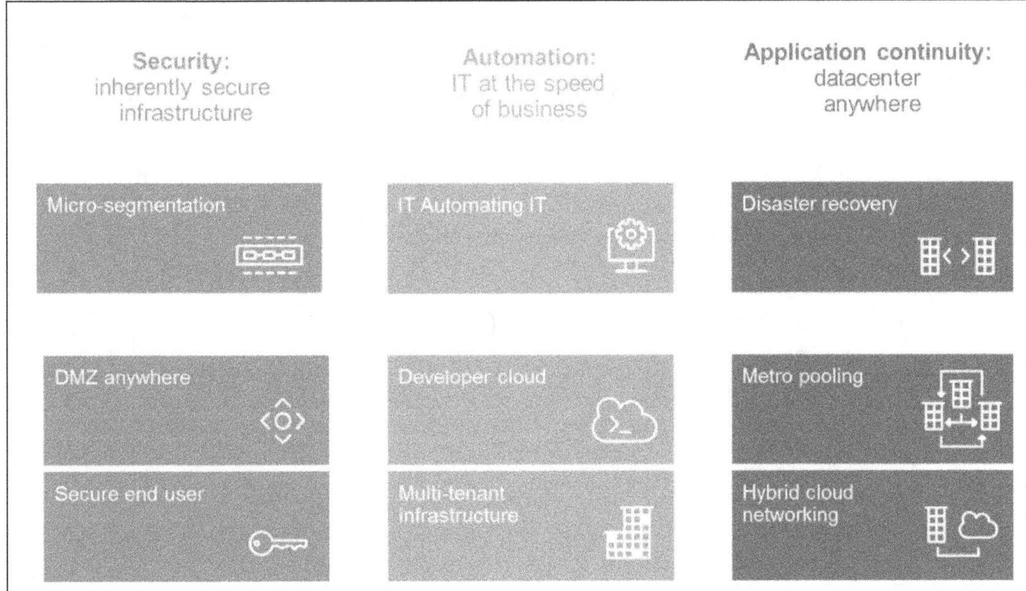

PowerCLI and NSX

As NSX is still fairly new in its wide scale deployment, many automation use cases have been requested from the user community to increase the accuracy and speed of deployment. One such use case was introduced through a VMware partner, with a staff member named Chris Wahl. His blog has been a significant source of virtual-based network information for many years. As a VMware User Group Champion, he posts and speaks at numerous locations.

Chris released a fantastic article and PowerShell script that is posted in his GitHub repository (reference: `https://github.com/WahlNetwork/nsx-tier-builder`). This uses a JSON file to gather data and then runs the script to create a typical configuration. Another VMware Technical Marketing individual (Brian Graf) has written a script to join an NSX Manager installation to a vCenter instance (reference: `https://github.com/vtagion/VMware-Products/blob/master/NSX%20Deploy.ps1`).

Both are fantastic script examples of what can be done with NSX and PowerCLI. However, if you look closely at either example, they don't utilize an NSX API call but use a web-based call to perform the actions. This really does provide limitless examples of what can be done with PowerCLI!

Orchestrating the integration of Site Recovery Manager

Site Recovery Manager (SRM) is a completely different type of tool from the NSX network platform. SRM is an orchestration tool that controls a storage array (SAN or NAS) or the vSphere Replication failover process. So, because it is an orchestration engine, if the failover process doesn't work manually, a SRM failover will not work either.

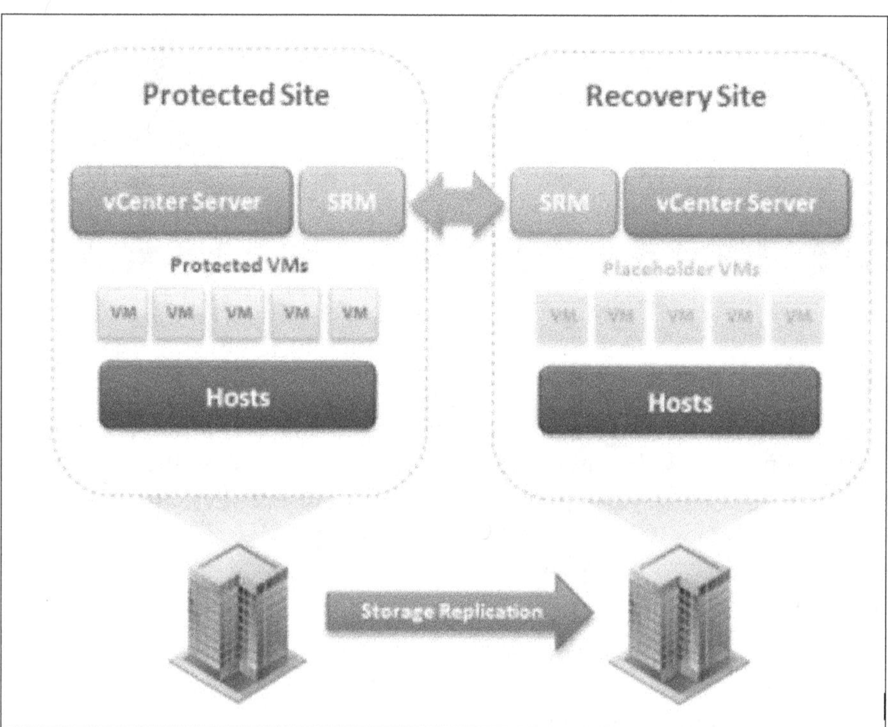

An SRM installation consists of a Windows installation where the SRM component is installed too, a database where the configuration is stored, and the vCenter registered extension or plug-in. In SRM 5.5 or older, it is integrated into the VIClient whereas SRM 5.8 and newer are integrated into the WebClient. Once the install is complete and the SRM install is connected to the vCenter, there are a number of things that PowerShell and PowerCLI can do.

PowerCLI and SRM

First, run the `Connect-viServer <vCenter>` command to connect the vCenter server.

Then, use `Connect-SrmServer` to connect to the SRM server that is linked to the vCenter server, or connect directly with `Connect-SrmServer -SrmServerAddress <SrmServer>`. Once connected, there are some components that can be scripted but not all. The public API allows the building and reporting of protection groups and recovery plans but doesn't control either the **Array Based Replication** (**ABR**) or **vSphere Replication** (**vR**). In some instances, storage vendors have public APIs that can be utilized to build, maintain, and destroy replications and they could be used if available but falls out of the scope of this book and chapter.

The VMware replication technology, vSphere Replication, does not have a public API at the time of writing this book and no scripting of individual VM replication can be performed because of it. So, the replication process has to be done through the Web Client or VIClient, depending on the version.

As with previous chapters, the use of an example helps in the process of understanding the integration. Busy Box Networks have noticed that their main office lies on a flood plain and the previous 2 years, the large river nearby has swelled and nearly overrun its banks causing road closures in the area of the warehouse. The company has debated moving to alternate locations but isn't feasible as other circumstances keep them in this location. They have also considered and had projects that have attempted to move their systems to a **Data Center as-a-service** (**DCaaS**) company not too far away. In the conversation with the ITaaS company, it is just going to be too expensive for what they need as they are happy with the level and expertise of the in-house staff. Then, the topic of **Disaster Recovery as-a-service (DRaaS)** was discussed and seemed to be the best of both scenarios and would save a tremendous amount of money as they wouldn't need to pay for equipment that isn't in use and keep the paid-for infrastructure in its place.

The project began and SRM was installed in the company and at the data center service, location, and sites were **Paired**. To save some time and effort vSphere Replication is chosen and a Layer 3 networking structure is put into place.

Layer 2 versus Layer 3 networking

In a site failover design, general considerations have to occur for a failover procedure. Is the IP address able to stay the same from one site to the other? If it is, then it is generally considered to have Layer 2 connectivity. The Layer concept comes from the OSI model (see reference here `https://en.wikipedia.org/wiki/OSI_model`) and presents a reference on how different machines are connected. **Layer 2** (**L2**) is defined as the Data Link Layer or the ability of a network device to appear within the same subnet or as if they are beside each other or on the same switch. L2 networking usually consists of very specialized equipment that fools the device into believing it is on the same network or it is using a network overlay technology as seen previously in the NSX section.

Layer 3 (**L3**) networking functions as a typical network would. It requires a routing device that is able to communicate through long distant connections. The Internet generally functions on a L3 communication plane. The largest issue is that an IP will have to change to accommodate the move between sites.

Each VM is designated as a candidate for failover and determined to be of one application, a Web server, an application server, and a database server. IP addresses are prepared and recorded within a spreadsheet, similar to that shown here, as the current address and the predetermined, non-used failover address:

VM	VM Purpose	IP Address	Failover IP
`Web01`	Web server for customer application	10.1.10.244	172.16.10.244
`App01`	App server for customer application	10.1.10.23	172.16.10.23
`DB01`	Database server for customer App	10.1.10.45	172.16.10.45
`DB02`	Database server	10.1.10.46	172.16.10.46
`Dom01`	Domain Controller	10.1.10.10	172.16.10.10

The three VMs `Web01`, `App01`, and `DB01` are replicated using vSphere Replication. Once they are configured, they can be added to a protection group and PowerCLI can be used:

```
$Cred = Get-Credential
Connect-VIServer VC01 -credential $cred
$srm = Connect-SrmServer
```

This establishes the connection to the linked SRM server and just like before, when $PS was used for remote PowerShell session, this can now be connected to the SRM API:

```
$srmAPI = $srm.ExtensionData
```

This link (`https://www.vmware.com/support/developer/srm-api/srm_61_api.pdf`) lists the API commands for SRM version 6.1 and just alters the digits in the URL shown to change to another version. With a connection to the API, it can be used to add VMs to a protection group, add mappings, or run a recovery plan. Additional API calls are shown in the PDF of the version of SRM that is being run.

On page 10 of the API documentation, it shows a heading labeled **Protection**, and under the heading, there's a command to call against it (`$srmAPI.protection.ListProtectionGroups()`), which will display available protection groups; and on page 12, `$srmAPI.recovery.ListPlans()` will display all the available recovery plans **Managed Object Reference (MoRef)**.

MoRef

There are a few reference posts in the VMware Blog regarding what a MoRef is. In short, it is an instance ID for an object in vSphere. For example, a VM has a MoRef that is a unique value for each VM. If there is a desire to see the MoRef for a VM, type `$(Get-VM vmname | Get-View).MoRef`.

If there is a need to see more information of the specific plans, run the command in a `ForEach` loop:

```
ForEach ($Plan in $srmAPI.recovery.ListPlans()) {
  $Plan.GetInfo()
  }
```

SRM is one of those applications that is typically a set-it-and-forget-it type of program; I personally, have found little use for scripting in this fashion. However, when daily tasks are listed for an operational team to check on the status of the associated protected VMs versus nonprotected, this mechanism can become important. As for Busy Box Networks, they have completed the replication and the creation of the three VM protection group and are now beginning on developing the Recovery Plan.

Running PowerShell scripts in a Recovery Plan

Busy Box Networks realize that the customer app requires some scripts to run for each of the components to be powered up in a certain order and after the Database server has started, a separate command needs to be run on both the Database servers and with the DNS before the App server and Web server start. Within a Recovery Plan, there is a location that allows a pre-startup or post-startup script to occur. Thinking this through, a pre-startup script cannot run on the affected VM and therefore must run on the SRM server itself.

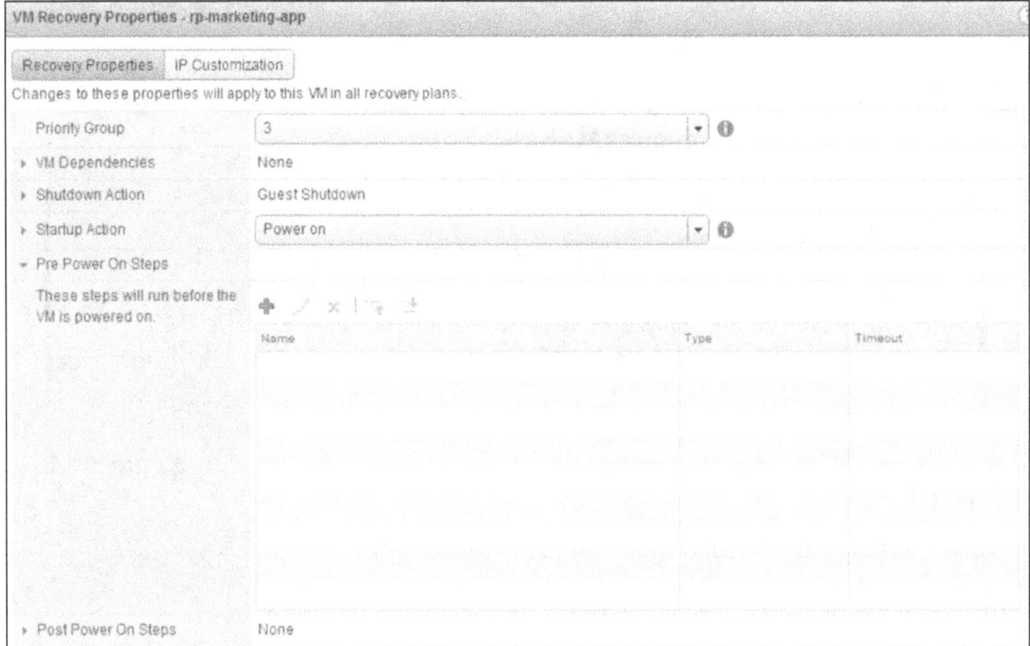

Expanding the **Pre Power On Steps** tab shows the information needed for an appropriate script. The **Pre Power On Steps** tab will allow the DNS entry to be edited for the VM before startup and runs from the SRM server itself.

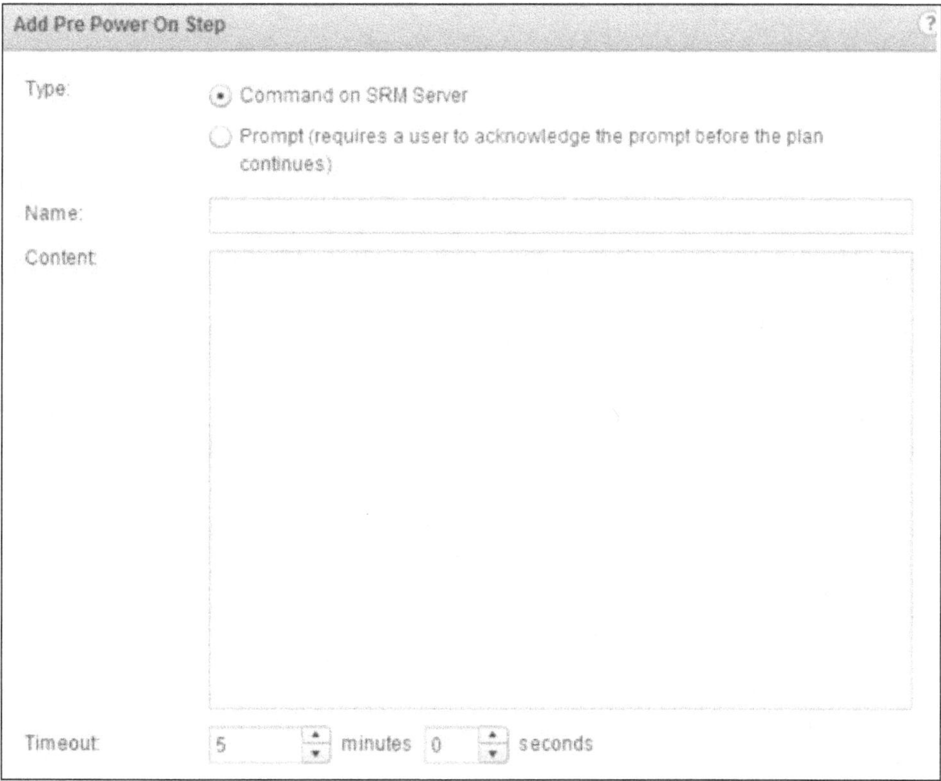

In the **Post Power On Steps** tab there is an additional option to run the command on the Recovered VM, for example, changing a database name or running a startup script for a service.

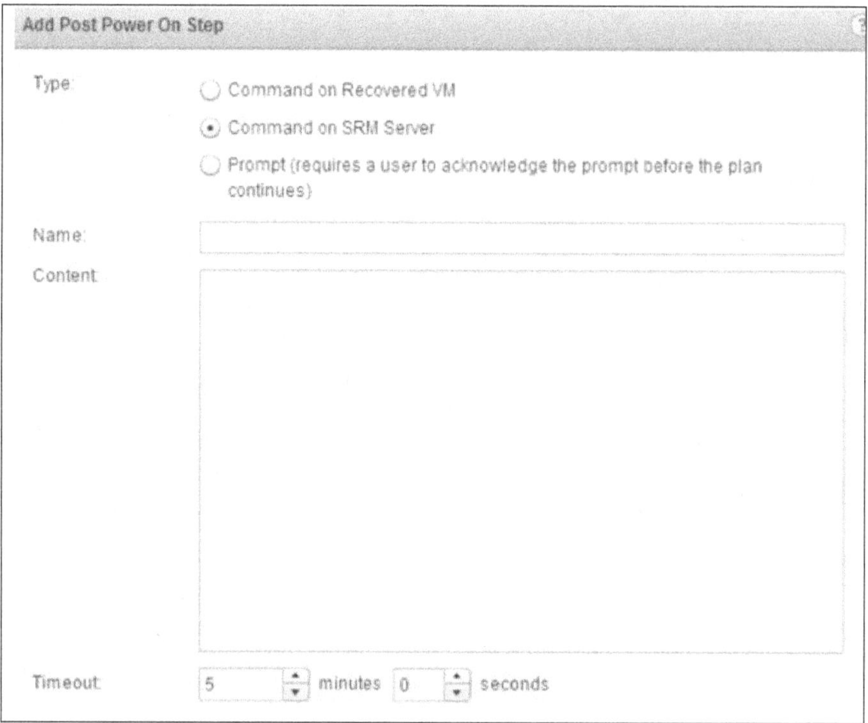

The **Name** field is a descriptor, whereas the **Content** field can contain an invoke-expression type of cmdlet or can actually contain all of the PowerShell commands in order. The key here is that it is just like hitting **Start** and **Run**, so the script must have the path and switches associated.

Orchestrating the integration of Monitoring (vROps)

vRealize Operations Manager provides key performance metrics to a complete Virtual Infrastructure Ecosystem, so it can connect to a single vCenter instance or to 50. It can poll and report on the NSX performance, VMware Horizon, or pull metrics from the NetApp or EMC array, or collaborate using **Log Insight** to categorize syslog data from a number of network devices. The tool itself is for viewing past performance and to help anticipate trends and future bottlenecks based on analysis of the data within.

Integrating PowerCLI and vROps requires a couple of specific versions to connect:

- **vRealize Operations Manager** (**vROps**) version 6.0.3 and newer
- PowerCLI at version 6.0 release 1 or newer

So if the environment that is being managed has older editions of vROps, there is no API to connect to.

Although there is a fairly straightforward install process, this chapter will only display the integration points of scripting and vROps. This section will explain some screens and concepts but assumes that the installation is complete, and has a single vROps Appliance in place. In vROps, the default dashboard upon login is the standby recommendations that display the **Health**, **Risk**, and **Efficiency** of the connected vCenter.

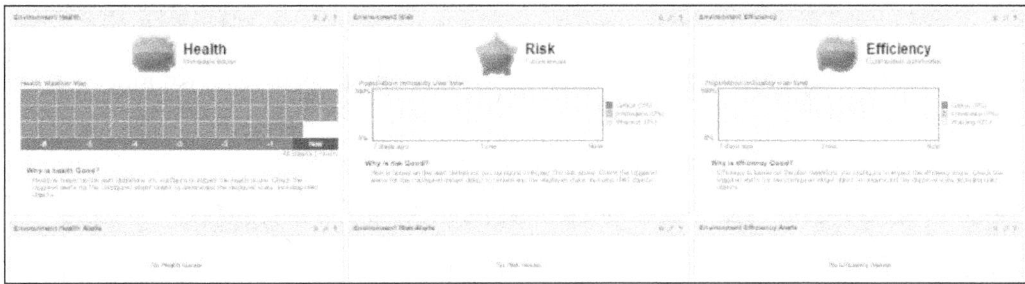

These are shown in a current configuration, future trends, and a present impact picture layout. The vCenter server and all the attached Hosts, VMs, and connected devices are categorized as objects. Using the idea that a host is an object, a VM running on the Host is referred as a child object. So, the recommendations screen displays the condition or status of the entire connected ecosystem and shows any alerts within the system. Selecting a specific object within the structure will show the relationships in regard to a parent or a child.

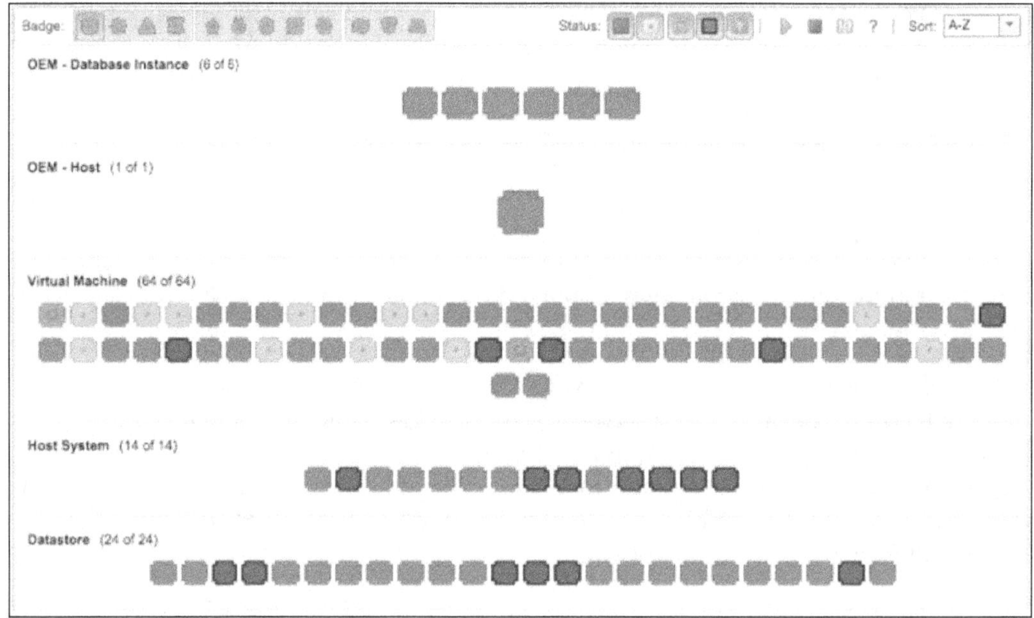

As the graphic shows **OEM – Host**, clicking on that object displays the connecting objects up or down the object inventory.

Connecting to the vRealize Operations Manager

Performing a connection to the server is done through the web interface, as a tab in the Web Client or through attaching through a PowerCLI interface.

The vROps web interface starts with this login screen:

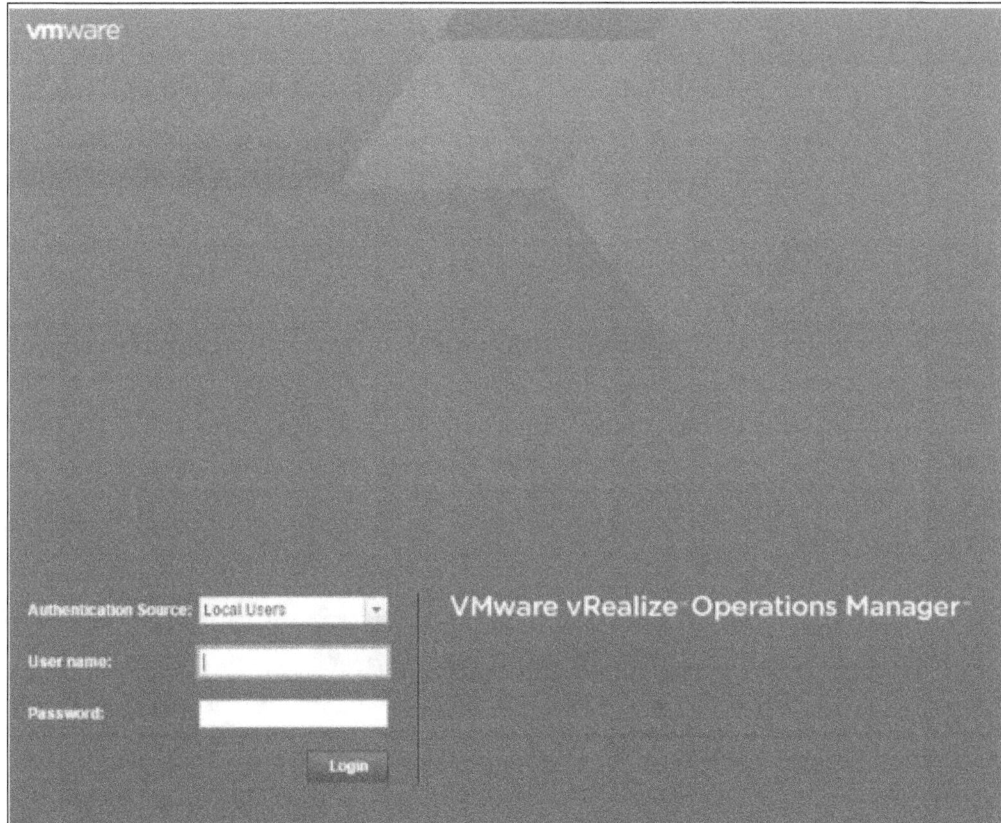

Then, it prompts for what is referred to as an authentication source. By default, the source is a local user database that contains a few essential accounts:

- Admin
- Maintenance Admin
- Migration Admin
- Automation Admin

The Admin account has full access and is the one used first to access the system. The first task on the system is to ensure that the system has a central authentication source that is not a local source. This will be critical later on if there is a desire to pass authentication during the connection to the vROps system. After the initial login with Admin, select the **Administration** link on the left-hand side of the window.

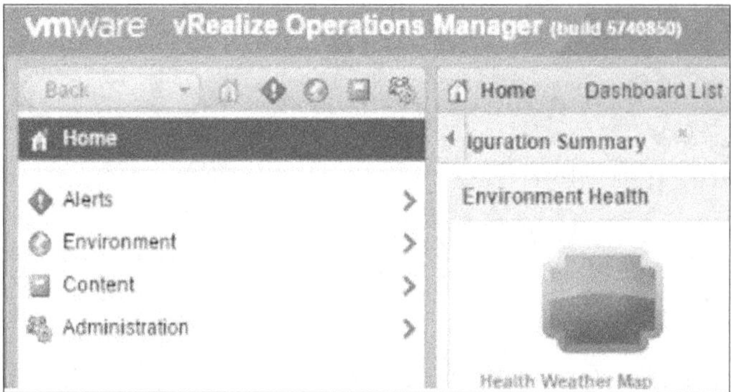

This expands the context and allows the Administration options to be displayed.

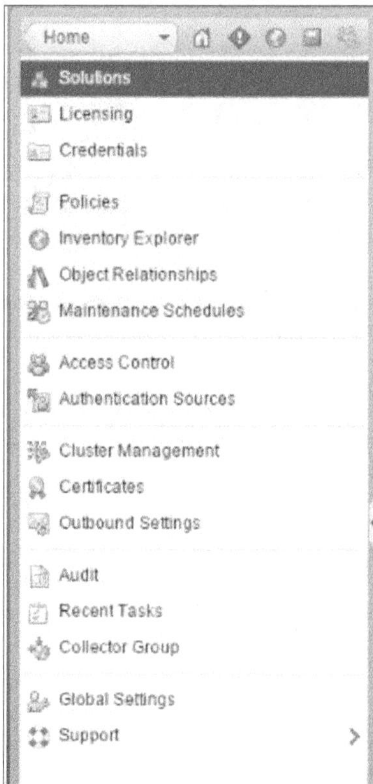

On a new build, **Authentication Sources** will be selected next in order to add the Active Directory or another LDAP connection that is to be created. Click on the **+** icon to add a new Source into the authentication schema. As of vROps version 6.1, there are four options for this area:

- SSO SAML
- Open LDAP
- Active Directory
- Other

The SSO SAML uses a connection to the **Platform Services Controller** (**PSC**) that is the SSO for vCenter when that was put in. The SSO can connect to whatever system has been entered into it, whether it is the default `@vSphere.local` or an attached Active Directory account. The PSC and SSO SAML are nice, in that, they can use any number of connections instead of specifying each one. Otherwise, the Open LDAP or AD connection can be used if there is only one AD domain.

Authentication Sources

Source Display Nar	Source Type ▲	Host	Port	Domain Name
Corp.local	Active Direct...	controlcente...	3268	corp.local
SSO	SSO SAML	psc-01a.cor...	443	vsphere.local

As seen, these are the examples of the authentication sources added to the system. This allows the import of users from a different source than local but doesn't mean any access has been granted. Select the **Access Control** option, which displays the user accounts. Move over one tab to the **User Groups** option and click on the Import Group icon. This is going to allow the import of an SSO-based group or an Active Directory Group into the system.

The import group wizard is started and a group can be chosen.

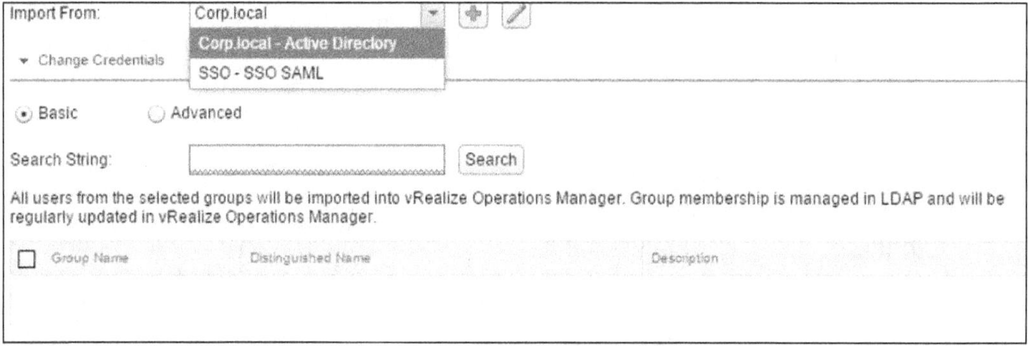

Once the group is selected, a role or level of permission is needed.

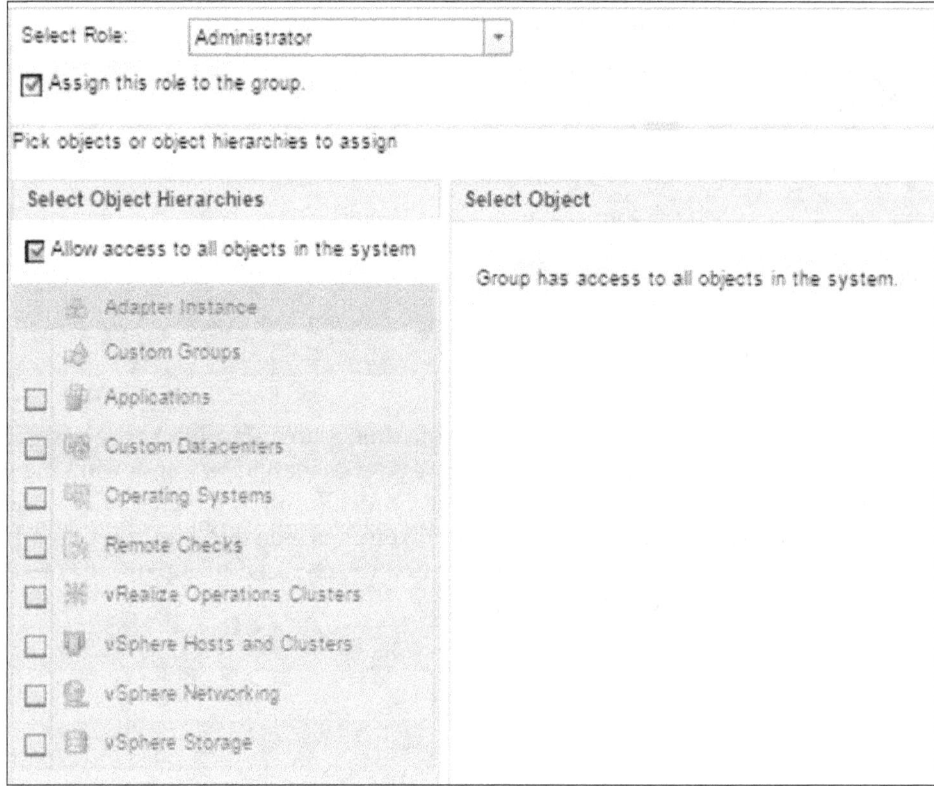

In this case, **Administrator** and all objects have been selected. After the import of the group permissions, the system will accept connections within PowerCLI.

PowerCLI and vROps

Similarly to SRM, PowerCLI is able to connect to the API associated to the application. The starting command is `vConnect-OMServer`, which, like SRM, has to be pointed to an object for easier script writing:

```
#Get the credentials
$cred = get-credentials
$OMServer = Connect-OMServer -authsource SSO -server vrops.corp.local -
credential $cred
```

Once that is connected, other OM cmdlets can be run. For example, setting a variable of `$haClusRes` to be an HA Cluster named `Cluster` can be set like this:

```
$haClusRes = Get-OMResource "Cluster"
```

And examining this object and the alerts that may be generated upon it could look like this:

```
$haAlert = Get-OMAlert -Resource $haClusRes -Status Active
```

This would display a result like this:

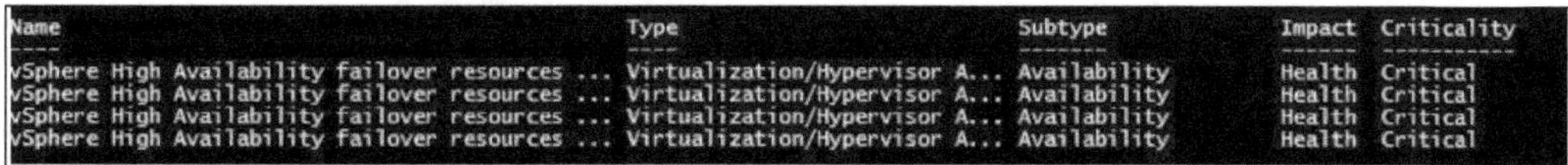

Name	Type	Subtype	Impact	Criticality
vSphere High Availability failover resources ...	Virtualization/Hypervisor A...	Availability	Health	Critical
vSphere High Availability failover resources ...	Virtualization/Hypervisor A...	Availability	Health	Critical
vSphere High Availability failover resources ...	Virtualization/Hypervisor A...	Availability	Health	Critical
vSphere High Availability failover resources ...	Virtualization/Hypervisor A...	Availability	Health	Critical

A similar example to the screenshot of the Web Interface is as follows:

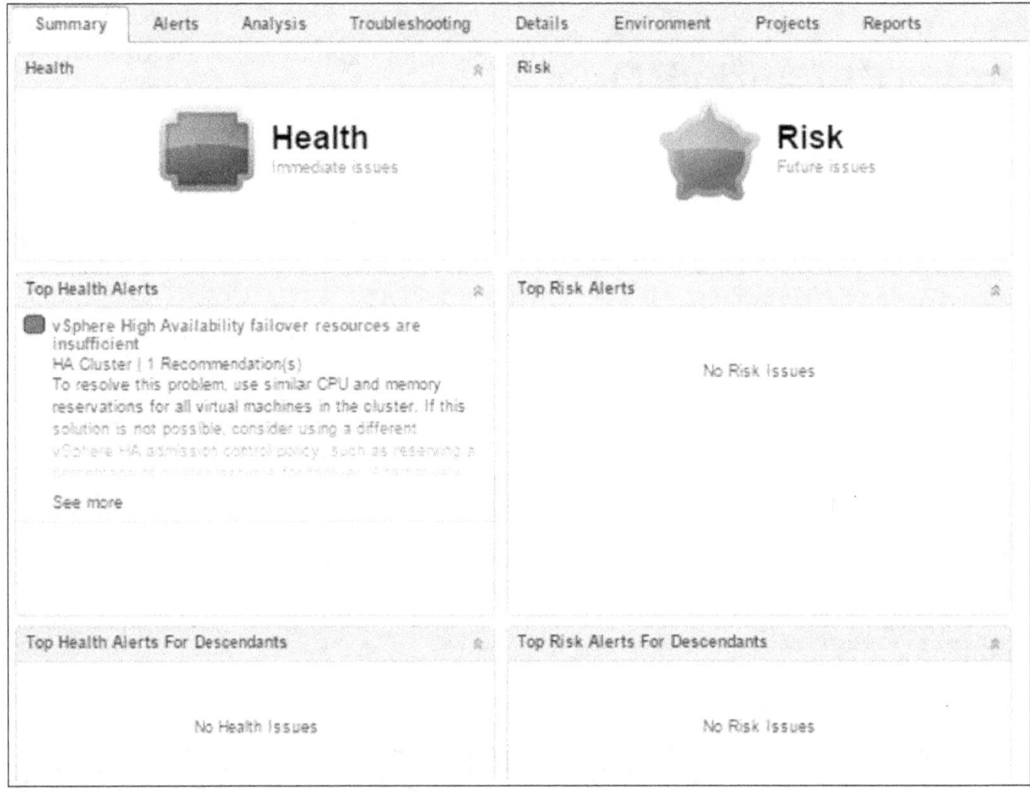

So, based on the alert, a second script can be run to alleviate the HA constraint on the cluster:

```
Get-Cluster "Cluster" | Get-VM | ? PowerState -match "On" `
| Move-VM -Destination $(Get-Cluster "Alternate Cluster")
```

There are a number of other components that can be explored in this area, such as recommendations, alerts, users, or resources. To get a listing of commands, type the following:

```
Get-Command -Module VMware.VimAutomation.vROps
```

Performance tools

As in the day of the life of any good VMware Admin, a painful user must fall. Of the countless environments I have personally experienced, there is always the question that "My VM is running slow". This generic comment solicits a blank stare from myself as I say "Define what slow is".

The common definition for those that ask is that the VM is running slower than it was the last time the user looked. It could be any number of things from a new update in Windows to a new application being installed on the VM. Thankfully, the performance tools within the VMware virtualization stack are excellent.

The three important commands are `Get-StatInterval`, `Get-StatType`, and `Get-Stat`. The `Get-StatInterval` command allows examination of the Statistics Sampling in the vCenter performance chart. The `Get-StatType` allows the ability to see what stats are available to the `Get-Stat` command where it retrieves the statistical information from vCenter.

Here are some examples:

- Run `Get-StatInterval`, and you'll get the following output:

```
Name                 Sampling Period Secs Storage Time Secs
----                 -------------------- -----------------
Past day             300                  86400
Past week            1800                 604800
Past month           7200                 2592000
Past year            86400                31536000
```

- Run `Get-VM "VM01" | Get-StatType`, and you'll get the following output:

```
cpu.usage.average
cpu.usagemhz.average
cpu.ready.summation
mem.usage.average
mem.swapinRate.average
mem.swapoutRate.average
mem.vmmemctl.average
mem.consumed.average
mem.overhead.average
disk.usage.average
disk.maxTotalLatency.latest
net.usage.average
sys.uptime.latest
disk.used.latest
disk.provisioned.latest
disk.unshared.latest
```

- Run `Get-Stat -Entity $(Get-VM vap) -CPU`, and you'll get the following output:

```
MetricId                Timestamp                         Value Unit     Instance
--------                ---------                         ----- ----     --------
cpu.usage.average       2/22/2016 4:00:00 PM               1.88 %
cpu.usage.average       2/21/2016 4:00:00 PM               1.83 %
cpu.usage.average       2/20/2016 4:00:00 PM               2.06 %
cpu.usage.average       2/19/2016 4:00:00 PM               2.21 %
cpu.usage.average       2/18/2016 4:00:00 PM                2.1 %
cpu.usage.average       2/17/2016 4:00:00 PM               1.92 %
cpu.usage.average       2/16/2016 4:00:00 PM               1.92 %
cpu.usage.average       2/15/2016 4:00:00 PM               1.89 %
cpu.usage.average       2/14/2016 4:00:00 PM               2.04 %
cpu.usage.average       2/13/2016 4:00:00 PM               2.63 %
```

- Run `Get-VM VM01 | Get-Stat -Common`

`Get-EsxTop` is the virtualization performance staple from a directly connected ESX Host, for example:

```
Connect-ViServer ESXiHost

Get-EsxTop -Counter
```

```
Name                Fields
----                ------
Server              {MinFetchIntervalInUsec:U64, IsVMVisor:B, TimeStampInUsec:U64, Time:S64}
PCPU                {NumOfLCPUs:U32, NumOfCores:U32, NumOfPackages:U32}
LCPU                {LCPUID:U32, CPUHz:U64, UsedTimeInUsec:U64, HaltTimeInUsec:U64...}
PMem                {PhysicalMemInKB:U32, COSMemInKB:U32, KernelManagedInKB:U32, NonkernelUsedInK...
NUMANode            {NodeID:U32, TotalInPages:U32, FreeInPages:U32}
Sched               {HostCPUInPct1Min:U32, HostCPUInPct5Min:U32, HostCPUInPct15Min:U32, NumOfSche...
SchedGroup          {GroupID:U32, GroupName:STR, IsValid:B, IsVM:B...}
CPUClient           {CPUClientID:U32, IsValid:B, NumOfVCPUs:U32}
HiddenWorld         {HiddenWorldID:U32, HiddenWorldName:STR}
VCPU                {VCPUID:U32, WorldName:STR, IsValid:B, AffinityStr:STR...}
VMem                {MemClientID:U32, IsValid:B, CurrentSwapInKB:U32, ToBeSwappedInKB:U32...}
VMNUMANodeMem       {NodeID:U32, IsValid:B, GuestMemInKB:U32, OverheadMemInKB:U32}
SCSI                {NumOfReservations:U64, DurationInUsec:U64, NumOfConflicts:U64, ConfigNumOfOu...
Adapter             {AdapterName:STR, IsValid:B, QueueDepth:U32}
Path                {PathName:STR, DeviceName:STR, IsValid:B, NumOfCommands:U64...}
WorldPerDev         {WorldID:U32, IsValid:B, NumOfActiveCmds:U32, NumOfQueuedCmds:U32...}
Partition           {PartitionID:U32, IsValid:B, NumOfCommands:U64, NumOfBlocksRead:U64...}
SCSIDevice          {DeviceName:STR, IsValid:B, QueueDepth:U32, BlockSizeInBytes:U32...}
Nfs                 {NumOfNfsClients:U32}
NfsClient           {MountName:STR, NumOfReads:U64, ReadByte:U64, ReadTimeInUsec:U64...}
Vscsi               {NumOfVscsiGroups:U32}
VscsiGroup          {GroupId:U32, NumOfVscsiDisks:U32}
VscsiDisk           {Name:STR, NumOfReads:U64, ReadByte:U64, LatencyReads:U64...}
Net                 {NumOfPortsets:U32, NumOfPNICs:U32}
NetPortset          {PortsetName:STR, IsValid:B, NumOfPorts:U32}
NetPort             {PortID:U32, IsValid:B, IsUplink:B, ClientName:STR...}
PNIC                {PNICName:STR, UplinkPort:U32, IsValid:B, IsLinkUp:B...}
Interrupt           {NumOfInterruptVectors:U32}
InterruptVector     {VectorID:S64, Devices:STR, NumOfCPUs:U32}
InterruptPerCPU     {CPUID:U32, Count:U64, SysTimeInUsec:S64}
Power               {NumOfLCPUs:U32, UsageNowInWatt:U32, UsageCapInWatt:U32}
CStateInfo          {StateID:S32}
PStateInfo          {StateID:S32}
TStateInfo          {StateID:S32}
LCPUPower           {LCPUID:U32, NumOfCStates:U32, NumOfPStates:U32, NumOfTStates:U32}
CState              {StateID:S32, ResidentTimeInUsec:S64}
PState              {StateID:S32, FrequencyInMhz:S32, ResidentTimeInUsec:S64}
TState              {StateID:S32, ResidentTimeInUsec:S64}
Vsan                {IsEnabled:B}
Dom                 {RoleName:STR, NumOfReadOps:U64, ReadByte:U64, SqSumReadLatencyInUsec:U64...}
```

Run `Get-EsxTop -TopologyInfo`, and you'll get the following output:

```
Topology          Entries
--------          -------
PCPU              {VMware.VimAutomation.ViCore.Impl.V1.EsxTop.DynamicDataImpl}
PMem              {VMware.VimAutomation.ViCore.Impl.V1.EsxTop.DynamicDataImpl}
CPUClient         {VMware.VimAutomation.ViCore.Impl.V1.EsxTop.DynamicDataImpl, VMware.VimAutoma...
HiddenWorld       {VMware.VimAutomation.ViCore.Impl.V1.EsxTop.DynamicDataImpl, VMware.VimAutoma...
SchedGroup        {VMware.VimAutomation.ViCore.Impl.V1.EsxTop.DynamicDataImpl, VMware.VimAutoma...
Sched             {VMware.VimAutomation.ViCore.Impl.V1.EsxTop.DynamicDataImpl}
Path              {VMware.VimAutomation.ViCore.Impl.V1.EsxTop.DynamicDataImpl, VMware.VimAutoma...
WorldPerDev       {VMware.VimAutomation.ViCore.Impl.V1.EsxTop.DynamicDataImpl, VMware.VimAutoma...
Partition         {VMware.VimAutomation.ViCore.Impl.V1.EsxTop.DynamicDataImpl, VMware.VimAutoma...
SCSIDevice        {VMware.VimAutomation.ViCore.Impl.V1.EsxTop.DynamicDataImpl, VMware.VimAutoma...
Adapter           {VMware.VimAutomation.ViCore.Impl.V1.EsxTop.DynamicDataImpl, VMware.VimAutoma...
SCSI              {VMware.VimAutomation.ViCore.Impl.V1.EsxTop.DynamicDataImpl}
NfsClient         {VMware.VimAutomation.ViCore.Impl.V1.EsxTop.DynamicDataImpl}
Nfs               {VMware.VimAutomation.ViCore.Impl.V1.EsxTop.DynamicDataImpl}
VscsiDisk         {VMware.VimAutomation.ViCore.Impl.V1.EsxTop.DynamicDataImpl, VMware.VimAutoma...
VscsiGroup        {VMware.VimAutomation.ViCore.Impl.V1.EsxTop.DynamicDataImpl, VMware.VimAutoma...
Vscsi             {VMware.VimAutomation.ViCore.Impl.V1.EsxTop.DynamicDataImpl}
NetPort           {VMware.VimAutomation.ViCore.Impl.V1.EsxTop.DynamicDataImpl, VMware.VimAutoma...
NetPortset        {VMware.VimAutomation.ViCore.Impl.V1.EsxTop.DynamicDataImpl, VMware.VimAutoma...
Net               {VMware.VimAutomation.ViCore.Impl.V1.EsxTop.DynamicDataImpl}
InterruptVector   {VMware.VimAutomation.ViCore.Impl.V1.EsxTop.DynamicDataImpl, VMware.VimAutoma...
Interrupt         {VMware.VimAutomation.ViCore.Impl.V1.EsxTop.DynamicDataImpl}
Power             {VMware.VimAutomation.ViCore.Impl.V1.EsxTop.DynamicDataImpl}
Server            {VMware.VimAutomation.ViCore.Impl.V1.EsxTop.DynamicDataImpl}
LCPUPower         {VMware.VimAutomation.ViCore.Impl.V1.EsxTop.DynamicDataImpl}
Vsan              {VMware.VimAutomation.ViCore.Impl.V1.EsxTop.DynamicDataImpl}
```

There are a number of additional ways to test and check the performance through testing configuration, and through checking deviations to that. Some of the scripts seen earlier in the book can attest to that as NTP, storage paths, DNS, and basic networking miscues will cause the issues. As with any solution, consistency is key to good performance.

Orchestration of Storage (VSAN)

Virtual Storage Area Network (**VSAN**) is a construct of locally attached disks assembled into a fully connected array of storage. VSAN is included into the kernel of later versions of vSphere (5.5 and newer) and allows the locally attached storage to be pooled and presented as a single datastore with the hosts connected to it.

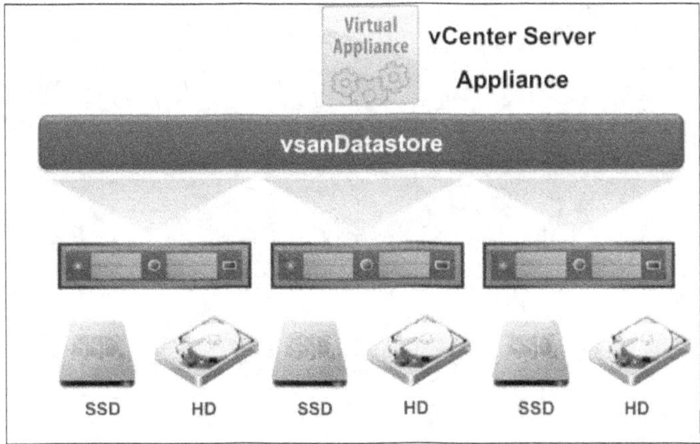

VSAN requires a cache tier as seen in the graphic as an SSD and a capacity tier as displayed as a HDD. This delivers fantastic performance and can be run on essentially commodity storage. This provides a huge cost benefit and gives the company a means to build a highly scalable solution for a remote office/branch office or running an entire infrastructure. It also simplifies the administration of the storage for the typical Virtual Infrastructure and gives an almost set-it-and-forget-it type of interface.

Setting up a VSAN is a fairly simple process as it has an attached disk to the host. Examining the **Hardware Compatibility List** (**HCL**) and locating the hardware and Disk Controller, which coincides to what is in the machine, allows a supportable configuration on the hosts. VSAN requires a minimum of three hosts to setup and run a datastore so as to allow a VMs VMDK or disk file to be in more than one place at a time for data integrity and continuity, so if an actual disk fails the data won't be gone.

Next is the setup of the network between the hosts. A VSAN network (a vmkernel network) is required to enable intercommunication of the data on all the hosts and disks. This must be a gigabit connection or higher due to the transfer of data between the machines.

Once those two pieces are running, licensing has to be applied to the last component. There are a number of additional factors that can be configured such as storage policy and the health analyzer but truly are secondary to the topic of automation.

VSAN and PowerCLI

VSAN has specific cmdlets that reside under the VMware.VimAutomation.Storage module.

```
CommandType     Name                              ModuleName
-----------     ----                              ----------
Cmdlet          Export-SpbmStoragePolicy          VMware.VimAutomation.storage
Cmdlet          Get-NfsUser                       VMware.VimAutomation.storage
Cmdlet          Get-SpbmCapability                VMware.VimAutomation.storage
Cmdlet          Get-SpbmCompatibleStorage         VMware.VimAutomation.storage
Cmdlet          Get-SpbmEntityConfiguration       VMware.VimAutomation.storage
Cmdlet          Get-SpbmStoragePolicy             VMware.VimAutomation.storage
Cmdlet          Get-VAIOFilter                    VMware.VimAutomation.storage
Cmdlet          Get-VasaProvider                  VMware.VimAutomation.storage
Cmdlet          Get-VasaStorageArray              VMware.VimAutomation.storage
Cmdlet          Get-VsanDisk                      VMware.VimAutomation.storage
Cmdlet          Get-VsanDiskGroup                 VMware.VimAutomation.storage
Cmdlet          Import-SpbmStoragePolicy          VMware.VimAutomation.storage
Cmdlet          New-NfsUser                       VMware.VimAutomation.storage
Cmdlet          New-SpbmRule                      VMware.VimAutomation.storage
Cmdlet          New-SpbmRuleSet                   VMware.VimAutomation.storage
Cmdlet          New-SpbmStoragePolicy             VMware.VimAutomation.storage
Cmdlet          New-VAIOFilter                    VMware.VimAutomation.storage
Cmdlet          New-VasaProvider                  VMware.VimAutomation.storage
Cmdlet          New-VsanDisk                      VMware.VimAutomation.storage
Cmdlet          New-VsanDiskGroup                 VMware.VimAutomation.storage
Cmdlet          Remove-NfsUser                    VMware.VimAutomation.storage
Cmdlet          Remove-SpbmStoragePolicy          VMware.VimAutomation.storage
Cmdlet          Remove-VAIOFilter                 VMware.VimAutomation.storage
Cmdlet          Remove-VasaProvider               VMware.VimAutomation.storage
Cmdlet          Remove-VsanDisk                   VMware.VimAutomation.storage
Cmdlet          Remove-VsanDiskGroup              VMware.VimAutomation.storage
Cmdlet          Set-NfsUser                       VMware.VimAutomation.storage
Cmdlet          Set-SpbmEntityConfiguration       VMware.VimAutomation.storage
Cmdlet          Set-SpbmStoragePolicy             VMware.VimAutomation.storage
Cmdlet          Set-VAIOFilter                    VMware.VimAutomation.storage
```

Some are VSAN-specific, such as `Get-VSANDisk`, whereas, others consist of `NFSuser` and VAIO and VASA configurations. As the majority of these cmdlets and use cases have not changed, a great VMware blog was the original one by *Alan Renouf* when this module was just a VMware Fling (reference: `https://blogs.vmware.com/PowerCLI/2013/11/vsan-and-vsphere-flash-read-cache-cmdlets.html`).

Summary

After numerous phone calls and e-mails attempting to find a time when Mr. Mitchell is available, his assistant responds with a positive meeting reply and you excitedly say, "He finally accepted!". After a number of vendor's requests, Mr. Mitchell accepts the request to tell the world about your IT transformation success.

"I am extremely pleased with the progress and what you have accomplished here." Mr. Mitchell says as he beams from ear to ear. "It has been an upward climb and you have come through it with flying colors, but I hate standing up at some vendor's conference to say that I was integral to the process. It was all the team, and I just said to go and do it." he says with a smirk.

Mr. Mitchell stands and leans over the boardroom table with an outstretched hand and says "I am glad to have you on my team, and we couldn't have done it without your leadership, insight, and tenacity to make this work."

I know this story is not what we always see in the IT job market and your boss may or may not extend this type of kudos, but understand that the practice of building DevOps is crucial to local IT shops keeping it relevant. Public cloud providers are making things cheaper and faster and soon keeping things local may not be in the best interest of the business.

This chapter dealt with trying to keep relevant and attempting to streamline the ability of the local IT in a dog-eat-dog type of industry. Automation, either through a tool such as PowerCLI or with a framework such as vRealize Orchestrator, or Automation allows us, as the Administrator, to keep pace with what the business demands.

Index

Q

R

S

T

V

W

www.ingramcontent.com/pod-product-compliance
Lightning Source LLC
LaVergne TN
LVHW081342050326
832903LV00024B/1265

* 9 7 8 1 7 8 5 8 8 1 7 7 0 *